Waste from Electrical and Electronic Equipment (WEEE) in Practice

Klaus Hieronymi

Waste from Electrical and Electronic Equipment (WEEE) in Practice

Manual for WEEE Implementation Based on Experiences

Springer

Klaus Hieronymi
Circular Economy Consulting
Oberursel (Taunus), Hessen, Germany

ISBN 978-3-031-40721-5 ISBN 978-3-031-40722-2 (eBook)
https://doi.org/10.1007/978-3-031-40722-2

This Springer imprint is published by the registered company Springer Nature Switzerland AG
The registered company address is: Gewerbestrasse 11, 6330 Cham, Switzerland

Paper in this product is recyclable.

Preface

In the early 1990s, the discussions of and about how using the Expanded Producer Principle as a basis to manage the growing E-Waste flows began and continues to this day.

This publication is based on experiences made by the author when participating in establishing operational solutions for E-Waste legislation in many countries across the globe and represent his personal view. Others may have different views on the developments. Specific legislation for E-Waste, very often called "WEEE-Legislation" (**W**aste from **E**lectrical and **E**lectronic **E**quipment) has been a new undertaking, countries experimented with various options at the beginning, some of them were discarded, others maintained, and new options invented.

Since the beginning of the discussion, one fact remains which should be considered regarding when discussing various options for E-Waste Legislation:

There is no "*One Size Fits All*" approach

The optimal design and implementation of EPR-based legislation and operation depends on many elements, most of them different from country to country, even within countries of a similar economic stage or within the same geography. This book can be used as a guide to select the appropriate options for the various elements to be considered for the respective legislative framework, its implementation and what elements may drive the effectiveness and efficiency of a local E-Waste System.

Please note that the prices/cost data used in this book are for example only, to illustrate the principle. Cost of collection/treatment is different in countries/regions, is very volatile, and can change within weeks.

This document intents to offer to all stakeholders involved in the development and implementation of regulation for "Waste from Electric and Electronic Equipment", from now on called **WEEE regulation**, to benefit from the experience made in many countries across the globe.

It focuses on legislation which is based on the principle of "Extended Producer Responsibility (EPR)" providing insights into options of various elements required when establishing, revising, and implementing, regulation in this area.

The document will especially outline the operational consequences of principles which are set in the design phase of E-Waste legislation.

As the countries with the most experience in WEEE legislation, are in the EU, many of the examples used in this document are from this region. However, in the meantime many countries in all parts of the world have experience with WEEE regulation and can provide valid information for successful WEEE implementation.

As no country is in the same situation, there is no "one size fits all" solution. However, there are certain principles which could ensure that legislation based on the Extended Producer Responsibility principle, called **EPR** legislation (not only for WEEE but also for packaging, batteries, etc.), establishes a level playfield for Producers, which allows (smart) Producers to gain competitive advantages on a fair basis. And finally, it help to establish an efficient Take-Back System (TBS) which provides maximum sustainability for the lowest costs (to all actors).

This document will not answer all questions linked to WEEE legislation/implementation. But it can provide some basic directions which can be adjusted to the situation in a specific country.

One of the challenges in writing this document has been to find the appropriate balance between overview and detail. WEEE is a complex subject, it contains environmental but also economic elements, process optimization, organizational setup, and psychology.

The majority of the information used in this book is from up to 2021 and may have been changed since then.

► All prices and other data used in this publication are fictive unless noted differently. They are very often accelerated to proof a principal rule.

Oberursel (Taunus), Germany Klaus Hieronymi

Acknowledgements

This book has been initiated by my previous colleges of the Environmental Department of HP Inc. who not only encouraged me to document my experiences in the development of WEEE legislation and their implementation across the globe. They also provided ideas and country information and acted as a sounding board to ensure that the content met the national realities.

I would also like to thank the members of the WEEE-Workgroups of the German Industry Associations for IT (BITKOM) and Consumer Electronics (ZVEI) as well as their European equivalent (Digital Europe) who allowed me to participate in developing the ideas which finally let to the WEEE-Directive and various implementations in the EU Member States.

Special thanks go to the founding members and the management of the European Recycling Platform (ERP), who enabled insights into processes, finance information as well as the specifics of countries not only in the EU. Namely Hans Korfmacher, who served as the founding CEO of this company, provided many ideas in the early days of the WEEE discussion which found their way in WEEE legislation across the globe.

I also thank representatives of NGOs, representatives of governments, associations, and PROs with whom I had (sometimes heated) discussions at conferences and within workgroups. These discussions were contributing to a framework of legislation and various implementation models which will make the world a little bit better place.

The Waste Management Departments of the cities of Kiel and Oberursel, both in Germany, allowed me to take photos from their E-Waste Collection facilities, many thanks for this.

A special "Thank You" goes to Matthias Schluep and his collegues from EMPA (Switzerland) for many debates around E-Waste and providing photos from the treatment of E-Waste in Ghana, India, and Egypt.

My thanks also to Prof. Dr. Martin Faulstich introduced me to academic principles of the E-Waste issue.

Last but not least, I would like to thank my family who supported my, e.g., thru literature research but also endured that I have been away for many meetings and conferences and may have not been the friendliest person when being stressed from, e.g., phone calls in the middle of the night.

Contents

Summary 1

Legislation to regulate *Waste from Electric and Electronic Equipment* (WEEE) has started in Europe in the early 1990s and since then spread around the globe.

Most of the laws and their implementations are based on the principle of *Extended Producer Responsibility* (EPR) which obliges Producers (=>those who put products onto the local market) to finance and (in most countries organize) the management of E-Waste.

The major principle of Extended Producer Responsibility (EPR) is to establish rewards for Producers to improve the Design for Recyclability (DFR) thru internalizing the cost of managing the post-consumer waste from their products.

The second goal, which has been the reason for a broad support of this principle among politicians, has been that any EPR-based waste management system immediately releases municipalities from waste management cost for the wastes concerned as they are moved from the municipality (=>citizen) to the Producer.

While ERP has clearly achieved to transfer of cost from municipalities to the Producers, rewarding DFR thru lower waste managing cost does only apply, if at all, to a very small extend in practice. An implementation of EPR based on *Individual Producer Responsibility* (IPR), where Producers care for their products is a very complex undertaking and needs the establishment of additional, costly processes (e.g., separation by brand, treatment of single brand batches). In conjunction with additional steps (e.g., how to deal with products from Producers who left the market) and the low recycling cost differences which can be achieved thru IPR today will lead to significantly higher cost. Other regulations and standards (e.g., RoHS, EPEAT) required design changes who are lowering recycling efforts (e.g., banning hazardous components which must be removed, to avoid contamination of recycling materials). The only country with an IPR-based Take-Back System for IT products is Japan.

K. Hieronymi, *Waste from Electrical and Electronic Equipment (WEEE) in Practice*,
https://doi.org/10.1007/978-3-031-40722-2_1

Collective Producer Responsibility (CPR), where Producers are made responsible for the entire waste (of their industry segment), is the basis of (almost) all E-Waste legislation across the globe.

Examples show that CPR does not reward automatically DFR, to the opposite, it provides a financial disadvantage to Producers who invest in increasing the recyclability of their products.

However, there are attempts in the EU (starting in France) to provide rewards for DFR by modifying the cost for waste management Producers are supposed to take by either *modulating* the fees or the amounts Producers have to collect/treat based on the attributes of their products. As these attributes are not limited to recycling related elements (e.g., repairability, warranty), the system is moving toward establishing an environmental bonus/malus system.

With the practical implementation of EPR systems, Producers will not only have to pay the cost of waste management, but they also have, in many countries, to set up structures to operate collection and proper waste treatment. In this document, various options for the elements of any WEEE regulation based on *Extended Producer Responsibility* are documented and analyzed regarding their impact with a focus on electronics. However, many of the elements described are valid for other waste streams as well (e.g., packaging, batteries).

2 Introduction

Extended Producer Responsibility (EPR) is an environmental protection strategy invented by Thomas Lundqvist, Prof. at Lund University (Sweden) in 1990 (Lindhqvist, 2000; Khetriwal et al., 2009).

EPR is based on the *Polluter Pays Principle* (PPP), established by the OECD first time in 1972 and implemented into the EU-Treaty in 1987 who establishes the financial responsibility of the polluter for all damages caused (Environmental policy: general principles and basic framework). Since then, this principle has been established in many national legislations around the globe.

Extended Producer Responsibility considers the Producer to be the polluter and consequently responsible for the environmental problems caused by the waste from the products he manufactured, irrespectively if the Producer has caused the specific environmental issues directly.

By making the Producer responsible for the entire life cycle of the product and especially for the costs of end-of-use management, investments in the *Design for Recycling* (DFR) should pay off due to lower cost of management end-of-life products.

Through practical implementations of the EPR principle, government wanted Producers to internalize waste management costs. Producers who designed their products to lower waste management costs (e.g., less hazardous materials, easy to dismantle) would be rewarded and those whose waste management costs are high would have a disadvantage.

Moving waste management costs to the Producers would also relieve the increasing waste management costs directly passed on to citizens (property tax). An effect which may have motivated even politicians who did not care much (at that time) for the environment to work toward EPR-based waste legislation.

K. Hieronymi, *Waste from Electrical and Electronic Equipment (WEEE) in Practice*,
https://doi.org/10.1007/978-3-031-40722-2_2

> **Remark** Waste management cost is paid by the final consumer anyway, independent if the waste is managed by municipalities (thru taxes) or by the Producers (included in the calculation of the sales price). Extended Producer Responsibility provides a chance, if properly implemented, to motivate Producers to lower the cost of waste management by optimizing the design of their products or waste operations.

There are many alternatives to set up the legal and operational system for managing waste according to EPR. In this document, the expression *Take-Back Systems* (TBS) includes all activities (e.g., obligation setting, collection, treatment) of all stakeholders involved, e.g., authorities, Producers, Dealers, Clearing Houses, and Producer Responsibility Organizations (PROs).

In the early days of EPR legislation, governments just wanted to move the costs of municipal waste management for a specific waste stream straight to the Producers but keep the operation with the municipalities; in principle establishing a new tax.

In Germany, the first country to establish legislation based on Extended Producer Responsibility in 1991 (on packaging), the Producers opposed the initial direction of just providing a financial contribution. Instead, Producers asked to operate the collection and treatment of the waste they were made responsible by themselves. Therefore, they established an organization to operate the management of packaging waste parallel (=>dual) to the municipal system. An organization called *Duales System Deutschland* (DSD) has been established which was responsible for the collection and treatment of packaging waste while the municipalities continued to manage all other waste (Fischer & Arndt, 2007). The logo of DSD has been the famous "Green Dot".

However, the experience with DSD has not been great for Producers. DSD became a practical monopoly as Producers had almost no alternative than to join DSD. As a result, waste management fees were very high, close to the costs of the packaging itself.

Due to pressure from big consumer companies, recyclers, and German/European antitrust authorities additional ***Producer Responsibility Organizations*** (PRO) were allowed to operate in 2003 (Kloiber, 2015).

As soon as the first competitor to the DSD joined the market, some Producers recognized that the fees they were supposed to pay for collecting and recycling packaging dropped dramatically (up to −70% in the following years).

In the 1980/1990s the consumption of electrical and electronic equipment (EEE) has grown exponentially, resulting in an increase of waste from electrical and electronic equipment (WEEE). According to the UN's Global E-waste Monitor of 2020, e-waste quantities rose by 21% within the past five years, making WEEE the fastest growing domestic waste stream by a large margin (Forti et al., 2020; Cucchiella, 2015). In particular, the content of hazardous substances and valuable resources makes the correct handling of EEE important. Improper handling leads to the endangering of human health and severe environmental degradation (Cesaro et al., 2017). Studies show that up to 70% of reported toxic and hazardous chemicals in the environment are caused by WEEE (Islam et al., 2020). At the same time proper treatment of WEEE bears immense economic potential and prevents the depletion of valuable resources, as raw materials from global WEEE generated in 2019 were valued at about $57 billion, but only a fraction of this value is extracted as part of waste management practices (Forti et al., 2020; Baldé et al., 2017).

To manage this waste stream, Sweden, Belgium, Switzerland and The Netherlands introduced EPR-based legislation already in the late 1990s for Electrical and Electronic Equipment (EEE). These national initiatives led finally to the first EU-Directive on "Waste from Electric and Electronic Equipment (WEEE)"(Directive 2003/108/EC) in 2003, revised in 2012 (Directive 2012/19/EU).

With WEEE considered to be the fastest growing waste stream, proper e-waste management shifted into the focus of governments and the electronics industry (Hieronymi, 2001).

Since the early 2000s countries across the globe established similar legislation, more or less following the EU-blueprint—making the Producers responsible for managing the end-of-use phase for their products. Several more countries are either in the process of implementing or revising existing legislation on WEEE management.

The existing national *Take-Back Systems* show significant differences in their approach, their level of regulation, and enforcement (Shittu, 2020).

The operational management of E-Waste obligations is organized and managed differently within the industry. Some Producers considered the cost caused by this new obligation as local sales cost while others considered it within their manufacturing/supply chain cost. Therefore, those who considered the WEEE-Cost as sales cost manage the obligation on a national basis. In the first days of this legislation, these Producers established *Producer Responsibility Organizations* (PRO) on a national basis. Those Producers, who considered the cost from their WEEE obligation as global manufacturing/supply chain cost soon identified the need for operational solutions on a Pan-European (later on a global) basis. Instead of dealing with multiple national PROs, these companies preferred to cooperate with PROs who operate on a pan-national basis. Amongst them have been Electrolux, HP, P&G and Sony who set up the first Pan-European PRO, the *European Recycling Platform* (ERP) in 2004.

As of today, Pan-European PROs, national and cooperation of national PROs are offering their services to Producers in Europe, some of them expanding their operations also on a global scope.

While the focus of early EPR-based legislation has been on managing significant increase of waste, minimization of pollution, and how it is financed, the emphasis has moved to resource conservation (with an emphasis on precious and Rare Earth metals) and lately waste minimization.

Another element fueling today's discussions about Electronic Waste (**E-Waste**) is the illegal transport of waste into developing countries where the scrap is treated not properly (e.g., burning PCs to get access to metals), many due to lower cost of inpropper recycling in these countries. With more advanced recycling technologies, the economy of scale, and the increase in transportation costs, prices for recycling electronics are, in many cases, today lower in industrialized countries, reducing the motivation for illegal exports of E-Waste.

The amount of E-Waste found in developing countries originating in the industrialized world is still significant. However, the major source is not illegally exported E-Waste. Most of the scrap results from the export of used products that are shipped "as is" without any sorting into products that can be Re-used and those which cannot. According to a report from the United Nations Environmental Program (UNEP), about 70% of the exports of used EEE to African countries can be Re-used without repair, 15% after a local repair, the rest are not re-usable and therefore E-Waste (Schluep et al., 2012). It can also be recognized that the amount of "*domestic E-Waste*" (E-Waste from products used in a developing country) is rising because of the market growth for new Electrical and Electronic Equipment in this area. However, national E-Waste legislation requires a strong focus to avoid the illegal export of e-waste. Establishing bans in E-Waste legislation is not sufficient unless loopholes are not closed (e.g., E-Waste declared as products for repair/re-use) and **legislation is properly enforced**.

In the EU, which in the past has been in the forefront of waste related legislation, a focus on extending the usage time of products is the latest trend in waste legislation. In conjunction with circular economy, legislation is looking into ways to encourage longer usage times with both, the users (e.g., lower sales tax on repairs (The Guardian, 2016) or Producers (e.g., an element of "modulation" (Fischer, 2017), see Sect. 3.1.7). Initiatives, e.g., in France intent to make the repairability of products more transparent to establish longevity as a criterium for the purchase decision (ECR Community, 2023). In 2019, the EU adopted Eco-Design implementing regulations within the Eco-Design-Workplan to improve, besides energy consumption, the repairability of electrical and electronic equipment (EU-Commission, 2019). While the market of secondhand products shows significant growth, re-using products from the waste stream is economically challenging. Cost of checking, repair and warranty are very often exceeding the potential sales price for these products which is limited due to the competition with low-end, no-name new products (Hieronymi, 2022).

While all E-Waste laws are based on the same principle, Extended Producer Responsibility, the details of the legislation and even more the operational implementation are not the same everywhere. The role of Producers, collection infrastructure, targets, services, number of PROs, convenience, and cost, just to name a few important elements differ to a large extend in the various countries. Even within the EU whose basis is the WEEE-Directive, details and implementation models are not the same in the EU-Member states (Cahill et al., 2011; Khetriwal et al., 2011).

Experience from countries who implemented legislation in the past can help those who either develop and implement E-Waste legislation or are revising existing legislation. There are a few publications who provide the academic basis and instruments for legislation. The "Waste electrical and electronic equipment

(WEEE) handbook" (Goodship et al., 2019) aims to pave a way for countries to pass legislation and form a coherent take-back system with lasting enforcement mechanisms. In its first chapters it offers very detailed assistance in the theoretical basis, structure, and enforcement of e-waste systems and based on the conclusions from existing systems, combined with the experience gained over many years and offers the academic basis for establishing E-Waste legislation.

If a country has decided to establish (or revise) it's E-Waste legislation, advice on details especially for the practical implementation is rare.

Questions often raised in conjunction with the implementation of E-Waste legislation are:

- What collection structure should be implemented?
- What type of Producer Responsibility is feasible?
- What is the role of Producer Responsibility Organizations (PROs)? How should they be set up?
- Should the collection targets be set by units or tons? Should they be fixed or relative to the amounts Put on to the Market (POM)?
- What products can/should be collected together or separately?
- What pricing model should be implemented?
- What is "Clearing" and when is it required?
- What is the impact of establishing multiple or single PROs?
- What should be the role of governments in the operation of a Take-Back System?
- How can a level playfield be organized to avoid unfair competitive advantages for some Producers?
- What are elements driving the cost of PROs and their efficiency?

Answers to these questions are important for, e.g., authorities, Producers, PROs, or other stakeholders.

References

Baldé, C. P., Forti, V., Gray, V., Kuehr, R., & Stegmann, P. (2017). *The global e-waste monitor–2017*. Bonn/Geneva/Vienna: United Nations University (UNU), International Telecommunication Union (ITU) and International Solid Waste Association.

BMUV. Bundesministerium für Umwelt, Naturschutz, Nukleare Sicherheit und Verbraucherschutz (2023).Verpackungsabfälle. Retrieved February 10, 2023, from https://www.bmuv.de/themen/wasser-ressourcen-abfall/kreislaufwirtschaft/abfallarten-abfallstroeme/verpackungsabfaelle

Cahill, R., Grimes, S. M., & Wilson, D. C. (2011). Extended Producer responsibility for packaging wastes and WEEE-a comparison of implementation and the role of local authorities across Europe. *Waste Management & Research, 29*(5), 455–479.

Cesaro, A., Belgiorno, V., Vaccari, M., Jandric, A., Chung, T. D., Dias, M. I., Hursthouse, A., & Salhofer, S. (2017). A device-specific prioritization strategy based on the potential for harm to human health in informal WEEE recycling. *Environmental Science and Pollution Research, 25*, 683–692.

Cucchiella, F., D'Adamo, I., Lenny Koh, S. C., & Rosa, P. (2015). Recycling of WEEEs: An economic assessment of present and future e-waste streams. *Renewable and Sustainable Energy Reviews, 51*, 263–272.

ECR Community. Implementing the Repairability Index in France. Retrieved February 10, 2023, from https://www.ecr-community.org/implementing-the-reparability-index-in-france/.

European Commission. (2019). The new Ecodesign measures explained. https://ec.europa.eu/commission/presscorner/detail/en/QANDA_19_5889. Visited Feb. 10th 2023

Fischer, T. (2017). *Modulated fees*. https://erp-recycling.org/wp-content/uploads/2017/11/ERP-Discussion-on-Modulated-Fees.pdf.

Fischer, K., & Arndt, H. W. (2007). *Kommentar zur Verpackungsverordnung* (Auflage 2). Verlag Recht und Wirtschaft.

Forti, V., Baldé, C. P., Kuehr, R., & Bel, G. (2020). The Global E-waste Monitor 2020. *United Nations University (UNU), International Telecommunication Union (ITU) & International Solid Waste Association (ISWA), Bonn/Geneva/Rotterdam.*

Goodship, V., Stevels, A., & Huisman, J. (Eds.). (2019). *Waste electrical and electronic equipment (WEEE) handbook*. Woodhead Publishing.

Hieronymi, K. (2001). Implementing the WEEE directive. In *Proceedings of the 2001 IEEE International Symposium on Electronics and the Environment. 2001 IEEE ISEE (Cat. No. 01CH37190)* (pp. 217–222). IEEE.

Hieronymi, K. (2022). Zur Ökonomie der Wiederverwendung. *Müll und Abfall*, 653–708.

Islam, A., Ahmed, T., Awual, R., Rahman, A., Sultana, M., Aziz, A. A., Monir, M. U., Teo, S. H., & Hasan, M. (2020). Advances in sustainable approaches to recover metals from e-waste-a review. *Journal of Cleaner Production, 244*, 118815.

Khetriwal, D. S., Kraeuchi, P., & Widmer, R. (2009). Producer responsibility for e-waste management: Key issues for consideration–learning from the Swiss experience. *Journal of Environmental Management, 90*(1), 153–165.

Khetriwal, D. S., Widmer, R., Kuehr, R., & Huisman, J. (2011). One WEEE, many species: Lessons from the European experience. *Waste Management & Research, 29*(9), 954–962.

Kloiber, B. (2015). Historie III: Der Grüne Punkt. Der Wertstoffblog. Retrieved December 8, 1999, from https://wertstoffblog.de/2015/10/29/historie-iii-die-dualen-systeme-die-kartellstrafe-fuer-dem-gruenen-punkt/.

Lindhqvist, T. (2000). *Extended Producer responsibility in cleaner production: Policy principle to promote environmental improvements of product systems*. Lund University.

Schluep, M., et al. (2012). Where are WEEE in Africa. In *Proceedings of Electronics Goes Green 2012+ (ECG 2012), Joint International Conference and Exhibition.*

Shittu, O. S., Williams, I. D., & Shaw, P. J. (2020). Global E-waste management: Can WEEE make a difference? A review of e-waste trends, legislation, contemporary issues and future challenges. *Waste Management*.

The Guardian. (2016). Retrieved February 10, 2023, from https://www.theguardian.com/world/2016/sep/19/waste-not-want-not-sweden-tax-breaks-repairs.

Important Elements of E-Waste Legislation/Implementation

3

3.1 Producer Responsibility

In practice, two operational alternatives of "Extended Producer Principle" are possible:

- Individual Producer Responsibility (IPR).
- Collective Producer Responsibility (CPR).

3.1.1 Individual Producer Responsibility (IPR)

Strict interpretation of the *Expanded Producer Responsibility Principle* (EPR) leads to an operational model where Producers are only responsible for the products they have produced (Individual Producer Responsibility, IPR). Therefore, in an IPR model, Producers get directly rewarded, if design changes on their products lead to lower waste management cost. IPR requires that products of different Producers are either collected separately or sorted by brand.

To the knowledge of the author, only Japan did implement a Take Back System based on Individual Producer Responsibility (IPR) for some IT-Equipment.

3.1.2 Collective Producer Responsibility (CPR)

The most common implementation of the EPR principle is *Collective Producer Responsibility* (CPR). With CPR, all Producers (e.g., of a certain product segment) are made responsible for managing "*their*" waste as a *Community of Producers* together. They are obliged to achieve the targets set by the government (e.g., for collection, recycling) and share burden of the collecting and treating operations

K. Hieronymi, *Waste from Electrical and Electronic Equipment (WEEE) in Practice*,
https://doi.org/10.1007/978-3-031-40722-2_3

they are responsible for as a collective. The cost is, in most cases, distributed between the Producers by market share, either by assigning a certain amount of the waste to a Producer (*Amounts Collected/Treated* (ACT)) or by a fee paid for each product sold (*Advanced Recycling Fee* (AFR)).

3.1.3 Individual or Collective Producer Responsibility?

Atasu et al. (2012) discusses in detail the pros and cons of both choices for implementing the EPR principle in E-Waste legislation. This paper analyzes the impact on the behavior of Producers regarding *Design for Recycling* (DFR) based on complex models including, e.g., market models, manufacturers perspective, recovery cost differentials, consumer surplus in both, IPR and CPR-based legislation.

In his conclusions, he states that "Producers find lower incentives for DFR under CPR than under IPR", a statement which many Producers are supporting (Dempsey et al., 2010).

In practice, Producers who implement DFR progressively do not only find lower incentives, but they even may also experience a competitive disadvantage with companies who ignore/invest less in DFR.

This issue is illustrated in Fig. 3.1. In the example, the market consists of (for simplicity) only 2 Producers, one who is investing into improving the recyclability of his products ("*Producer 1*"). He may use, e.g., screws instead of glue to allow easy dismantling or replacing materials containing hazardous substances (e.g., certain flame retardants). It is assumed that the additional cost to develop and implement the design changes is $1/product sold.

	Products sold	DFR Investment/ Product	Total DFR Investment	Take-Back Obligation (50% Collection Rate)	Products with DFR in Collection
Producer 1	10,000 units	$1	$10,000	5,000 units	**500 units**
Producer 2	90,000 units	$0	$0	45,000 units	**4,500 units**

Fig. 3.1 Investments in DFR can lead to higher manufacturing cost

"*Producer 2*" is focusing on reducing production cost and does not take DFR into consideration.

In this example, *Producer 1* will face additional manufacturing cost due to adopting DFR Criteria of $1/unit. For his entire production of (assumed) 10,000 units, his investment in DFR will sum up to $10 k in total. *Producer 2*, who does not change the design of his products will not experience any additional manufacturing cost for the 90.000 units he produces.

Assuming further that the overall Collection Rate for products of *Producer 1* as well as *Producer 2* is 50%, 50.000 units will be collected in total. As a consequence 5.000 units of *Producer 1* (with DFR) and 45.000 units of *Producer 2* will be found in the collected waste.

In a CPR-based legislation where all Producers share the responsibility for all products, *Producer 1* (market share 10%) will be responsible for managing 10% of the products collected (5.000 units) while *Producer 2* will manage 45,000 units resulting out of his market share of 90%.

Considering further that not intended or unintended sorting according to brands will be applied, the amounts *Producer 1* is responsible for (5.000 units) will contain 500 units of his own products with DFR attributes, and the remaining 4.500 products in his obligation have been manufactured by his competitor without DFR.

In today's practice, recycling cost savings of products with similar technology are quite small, maximum a few cents/unit. To illustrate the missing rewards for DFR in a WEEE-System based on Collective Producer Responsibility, the assumed saving have been set to $2/unit, far above realistic expectations.

As shown in Fig. 3.2, recycling cost savings of $2/unit from the 500 units with DFR found in the obligation of Producer 1 result in cost savings of Producer 1 of $1,000. Compared to his investment of $10,000 in the production of recycling friendly products he will experience a negative ROI. Producer 2 however, will benefit from the DFR-investments of his competitor as 4.500 units in the waste he is responsible for are products of Producer 1 who carry lower recycling cost.

	Recycling costs reduction of products with DFR	Recycling costs savings (Products w DFR in Collection * Recycling costs reduction / Unit)	Recycling costs saving – DFR Investment
Producer 1	$2/unit	5oo units * $2 = $1,000	$-9,000
Producer 2		4,500 units * $2 = $9,000	$9,000

Fig. 3.2 ROI of investments in DFR in a WEEE system with collective Producer responsibility

Producer 2, who did not invest in DFR would face significant cost savings as the amount he is responsible for will contain 4,500 units with DFR resulting in lower cost of $9,000.

> **Conclusion** In a WEEE-System, based on Collective Producer Responsibility (CPR), those Producers who care about DFR will not be rewarded. To the opposite, those who do not invest in DFR benefit from the investments of those whose products are easier to recycle."Modulation" (see Sect. 3.2.5) may cross this gap.

If CPR is not rewarding DFR, why is, with the exception of PC's Japan, no EPR legislation for E-Waste based on IPR?

The alternative model to CPR is the Individual Producer Responsibility (IPR). IPR is a great academic principle to reward Producers, who invest in DFR, practical implementation is very difficult and costly.

To achieve the recycling cost savings for those Producers who invest in DFR, products must be processed separately for every individual Producer, e.g., to gain the recycling cost advantages, specific recycling processes must be applied for DFR-products and for the products who where DFR-principles have not been followed. In practice, this requires that products of a Producer who invested in DFR must be separated from those who do not care. Separation can be done by either collecting products separately by brand or sorting the waste before processing.

Differentiating between brands at collection requires significantly more space and equipment (e.g., containers) at collection sites or different collection trucks with curb-side collection. Even if enough space is available at collection sites, separating products by brand within, e.g., the "*screen category*" with>50 different Producers is impossible in practice.

In addition, IPR may require treatment of the waste in batches, as no recycler will receive enough waste to feed a brand-specific recycling line (unless waste is transported over huge distances to central processing sites) on a permanent basis.

Separate collection and processing by Producer will significantly increase the cost of waste management. From an economic principle, waste management cost of products with DFR would have to be significantly lower than the additional manufacturing, collection, and recycling cost.

▶ **Remark** Collection and treatment costs of Take-Back Systems based on Individual Producer Responsibility are, due to separation and treatment in separate, brand-based batches usually exceeding the cost of those based on Collective Producer Responsibility. The reality contradicts the academic principle.

Since RoHS, an EU-legislation to reduce hazardous substances in EEE, has been established in 2002 and revised in 2011 (Directive 2002/95), the amount of problematic materials, a major cause of high recycling cost differences has been dramatically reduced. This regulation, even enforced only in the EU, became a standard for almost all products across the globe as manufacturers apply the same design rules for their entire product spectrum and do not differentiate their design for different regions/countries.

EPEAT (Electronic Product Environmental Assessment Tool, Global Electronics Council, 2015) is an Eco-Label, which sets environmental criteria for the procurement of the US government, it is also used by other governments, and large cooperations across the globe. Most Producers in the IT-Segment, are following the DFR-criteria set by this label. EPEAT contains various DFR elements, another reason why DFR has been widely increased across many products of the IT industry in the past years across the industry.

In conjunction with innovations in the recycling industry (e.g., improvements in sorting plastics), the opportunities to significantly reduce the recycling cost thru DFR for electronics are very small.

For example, for a current Notebook, the difference between the recycling cost of a product where DFR is maximized and those who just follow legal obligations/ Eco-Labels (e.g., from ROHS, EPEAT) is as low as a few US$-cents/unit.

The additional efforts in logistics (individual pick up) in Japan, the only E-Waste legislation based on IPR (for PC & Monitors), lead to collection and treatment cost (per unit collected) which are>50 times higher as charged by PROs in countries with E-Waste legislation based on Collective Producer Responsibility.

▶ **Conclusion** In the current framework, IPR does not deliver significant rewards for Producers who invest more in DFR.

Another issue in an IPR-based EPR legislations is *Orphan Products*. Products who have been put on to the market by Producers who left the market (e.g., stopped selling, bankruptcy) before their products became waste.

To avoid the situation that there is no Producer available who takes the responsibility for these products, in a TBS based on IPR, individual Producers would have to ensure that funds are available in case they leave the market. According to International Financial Reporting Standards (IFRS) and the Generally Accepted Accounting Principles (GAAP), accruals must be established in each Producer's balance (sheet) for future liabilities of their current business (Walther, 2013).

These accounting rules would require that each Producer would have to set up accruals to cover the collection and recycling obligation under a WEEE-type legislation based on IPR for each product sold today, for when they are turned into waste in the future.

The obligations resulting from E-Waste legislation in conjunction with IFRS/ GAAP could result in the need to establish huge accruals for all products a Producer sold in the past years (=>Historic Waste). Especially in the first year, when accruals must be established for the entire installed base, these accruals can severely impact the annual results of Producers.

Even when Producers active on the market have established accruals, one problem remains unsolved: Who takes care for the waste of *Orphan Products*, products from those Producers who left the market and their products will be returned in many years to come?

In case there is no funds available to finance the waste from Producers who left the market, one option is that customers must pay a fee when they dispose a product, a practice which is established for Large Domestic Appliances in Japan (e.g., for LDA). Or the *public*, e.g., the municipalities would continue to finance the *Orphan Products* and historic waste.

In a TBS based on Collective Producer Responsibility (CPR) the management of *Orphan Products*, historic waste does not require any accruals either with Producers or with PROs.

Under CPR, all Producers are responsible for waste all waste returned relative to the amount of products sold in the same period, independent if the returns consist out of *Orphan Products* or *Historic Waste*.

Making existing Producers responsible for the waste from another brand is rational in conjunction with the "Polluter Pays Principle". Most of the products sold (at least in mature markets) are replacing an existing product. The seller of the new item could be considered as the *Polluter* as his product is the cause of retiring the old one.

> **Conclusion** In WEEE-Systems, based on Collective Producer Responsibility, the Producers active on the market today pay for all the waste from retired products, independent if their Producers have already left the market or are still active.

In principle, Producers are made responsible for a fraction of the e-waste generated in relation to the amounts they put onto the market. If, e.g., a Producer's market share (in a product segment) is 25%, he is responsible for 25% of the waste arising in that sector, independent of the brand of the products or when they have been sold.

As in a CPR-based system all waste, independent of who has caused it, will be taken care by the Producers active on the market, Producers who leave the market do not have any liabilities for the products still in use and therefore are not obliged to establish accruals for its future waste.

> **Remark** A CPR-based WEEE-System will avoid any accruals inside Producers to cover the waste management cost of their products in the future. *Historic Waste* and *Orphan Products* are taken care of as the waste from these products is assigned to existing Producers. In a nutshell, WEEE-Systems based on Collective Producer Responsibility (CPR) are much easier and less complex to implement.

However, a CPR-based system carries several risks of cross subsidizing/ blending WEEE cost between individual Producers, product groups, and or waste categories. For details see Sect. 4.2.2.

Rewarding DFR is also possible in CPR-based systems, but not without additional effort. For example can Producers establish brand related "Take-Back Events", where they could compensate the additional costs of individual collection, for example, by using them as marketing events (=>old for new). Such events have been a proven marketing tool, especially when combine with a raffle or competing elements. For example, in Frankfurt/Germany, a Producer organized a competition of sports-clubs. The club who collected most retired IT-Products received a (new) laptop, the 2nd a Laserprinter and the third an Inkjet printer. Total collection has been more than 250 units within 4 h.

Communicating that the devices collected will be checked by a social institution for re-use will also be an excellent motivator for consumers to bring back old equipment. At event in France, more than 300 devices were collected within a day, and 35 have been considered to be able for re-use.

Differentiating the obligation of Producers based on the environmental footprint (*Modulation*) is another method to (indirectly) reward DFR. Details can be found in Sect. 3.2.5).

3.1.4 Goals/Targets

The objective of any WEEE legislation around the globe is to avoid any environmental impact (e.g., pollution through burning or illegal dumping) of electronic and electrical equipment when discarded at the end of its useful life as mentioned in the EU-WEEE-Directive:

> "...preventive action should be taken, that environmental damage should, as a priority, be rectified at source and that the polluter should pay" (2012/19/EU)

With the awareness on the potential shortages of mineral resources, the recovery of materials, especially metals have been added to the goals E-Waste legislation should achieve.

▶ The ultimate measure to rectify environmental damage from E-Waste is that no waste is dumped or treated illegally and all waste management is paid by the polluter (=> Producer).

As establishing and measuring targets for these goals is quite difficult (e.g., how to measure illegal dumping?), most countries established targets for the collection of E-waste.

In first WEEE-Implementations, these targets were set as an amount of e-waste collected per capita (e.g., in the EU 4 kg/inhabitant), later as an amount in relation to products sold. All these measures have advantages on one side, but also difficulties due to incomplete data and the complexity of the various e-waste streams. In parallel, targets regarding recycling have been set, usually % of materials recycled vs. the total amount treated.

Caused by the latest discussions regarding the extension of usage time of products, some countries are about to adopt targets to measure and increase the number of products re-used from waste. All Targets in the WEEE-context can be based on either units (pieces, devices) or weight.

3.1.4.1 Units (Pieces) as a basis of Targets

The number of units (e.g., Put on Market or collected/recycled) was a very common measure to set and control targets in the early days of WEEE. However, most countries have moved away from this form of measurement after recognizing that unit measurement leans toward an unfair distribution of E-Waste responsibility between Producers of products with different size/weight.

Transportation and recycling costs are usually managed/charged by weight. In this context, a generic allocation of the Producer's obligations/costs by the number of devices would benefit the Producers of heavy products as they are required to provide the same contribution per unit as those who sell lighter devices.

While the number of products Put on the Market is quite easy to establish by the Producers, the number of units collected would include an additional, costly step: counting products in the e-waste collected (what about products who broke apart?), either at Collection Sites or before recycling takes place.

3.1.4.2 Weight as a basis of Targets

Weight is the most important driver of collection/transportation cost. Companies who transport waste charge by weight (and distance), recyclers as well and incineration/disposal fees include also a weight element. Therefore, the data of the weight of retired equipment is available without further efforts.

3.1.4.3 Collection Targets

Collection Targets are an essential element to achieve the goals of WEEE-Legislation: All End-of-Use Products are collected and receive proper treatment; materials are re-used and the cost is taken care by the Producers.

Targets who are set to low will cause waste to be *left behind*; those not collected may end up in improper treatment or are even discarded outdoors or the cost may end up with municipalities.

Targets aimed to high may raise costs significantly without a positive impact on the amounts collected. If the collection target is higher than the available E-Waste, people may store E-waste and sell it to the bidder with the highest offer.

> **Remark** Setting realistic Collection targets is an important element for the success of E-Waste legislation and the competitiveness of national Producers.

Two different options for setting targets can be observed in various countries:

- Fixed Collection Targets.
- Flexible Collection Targets.

3.1.4.4 Fixed Collection Targets

Some countries are establishing a target out of a fixed amount of waste to be collected by Producers. The target, sometimes set specifically by collection group are either based on research (e.g., UK) or just on "*Guestimation*". Very often they are a result of negotiations between authorities and Producer associations.

Some examples of *Fixed Targets*

- New York: ~100 M lbs. for all Producers (individual manufacturer goal determined by market share).
- New Jersey: 51 M lbs. for all Producers.
- Wisconsin: ~22 M lbs. for all Producers.

The challenges of this method are:

- Determining realistic goals is difficult. If the targets exceed the amount of waste generated, E-Waste collected by commercial actors is sold to Producers who need it for their compliance at very high prices. If it is too low, not all E-Waste will be collected and treated properly. Actors may just pick up e-waste with positive recycling value and may leave the rest.
- Adjusting fixed goals to changes in the market (e.g., more products sold, technical breakthrough causes the retirement of more products than expected) is quite difficult and time consuming.
- Commercial actors manage the valuable equipment (e.g., Notebooks) on their own and only provide those with high treatment cost to Producers. Consequently, IT Producers must buy CRT-based TVs to fulfill their obligation in some US-States.
- Some Producers established alternative collection routes (e.g., in the US through dealers), but they never reached amounts to fulfill their obligation completely.

Australia applies a method to set the Collection Target considering elements of *WEEE-Generated*. As there are no major OEMs producing electronic equipment locally, they take the import data, deduct the amounts that have been officially exported and consider that about 30% of the waste is leaked (e.g., exported as a second-hand product, put in household waste). Therefore 70% of the net imports (of new products) are considered to be *Waste Generated* and should be collected by the Producers.

3.1.4.5 Flexible Collection Targets

Other than Fixed Targets, Flexible Targets adjust automatically to changes in the market. For example, if more E-Waste is generated, the obligation of Producers will raise automatically.

The most common Flexible Target used is the Collection Rate.

The Collection Rate (*CR*) is defined as the relation between *Waste Properly Collected and Treated* (*WCT*) and all *E-Waste Generated* (*WG*) of a certain market in a given year (n).

While the calculation of CR according to the formula in Fig. 3.3 appears to be simple, the challenge is to get the data required. Especially data for *Waste Generated* (WG) as well as *Waste Properly Collected and Treated* (WCT) might be not available.

$$CR_n = \frac{WCT_{(n)}}{WG_{(n)}}$$

Fig. 3.3 Basic formula to calculate the Collection Rate (CR)

E-Waste Collected and Properly Treated (WCT)

Measuring the amount of E-Waste properly treated (sometimes called "*WEEE Returned*") is a complex undertaking. An exact measurement may require, if it is possible at all, significant additional reporting, and controls.

Many E-Waste legislation implementations were using the amounts collected by the Producer Responsibility Organizations (PROs) to measure the Collection Rate. This is underestimating the real amounts collected/treated as many E-Waste flows outside the PRO are also ending at proper recycling facilities and therefore must be considered as Collected and Properly Treated (see Fig. 3.5).

E-Waste is collected and managed outside of the PROs out of commercial reasons. Net recycling values from IT and large Domestic Appliances can reach up to up to $5/unit (Washing Machine), they heavily depend on scrap prices. Recycling revenues of different products are captured in Fig. 3.4, based on a survey of recyclers in Germany in 2019.

Material Value of various Products

	Laptop	Desktop (old)*	Tablet	Smartphone	Inkjet Printer	Laser Printer
Polymers	$0.05	$0.09	$0.01	$0.00	$0.19	$0.38
Glass/Ceramics	-	-	-	-	-	-
Ferrous metals	$0.24	$1.11	$0.10	$0.03	$0.41	$2.28
Precious metals	$8.68	$18.04	$5.16	$1.13	$1.72	$2.64
Rare metals	$1.51	$0.84	$0.07	$0.02	$0.01	$0.00
Other	-	-	-	-	-	-
Material value**	$11,64	$24.11	$5.67	$1.26	$2,76	$6.56
Achievable Recycling value**	$1–3	$2–5	$0–1	$0.5–1	$0.0	$2–3

* Desktops produced before 2010. Newer Products contain significantly fewer precious metals
** Net recycling values, depending on amounts treated, recycling technology etc.

Fig. 3.4 Value of materials of products at production. Not all materials, especially those who are used in very small quantities can be recovered (e.g., Rare Earth Metals)

Very often, individuals but also small companies (=>*Informal Sector*) are collecting retired products with a positive recycling value and sell them to recyclers. But also, retailers, social enterprises, and even municipalities are managing the valuable fractions of waste on their own, not providing them to the PROs.

Huisman et al. (2012) have analyzed the various flows for The Netherlands, their results are documented in Fig. 3.5. Similar studies in other countries provide similar trends/data points.

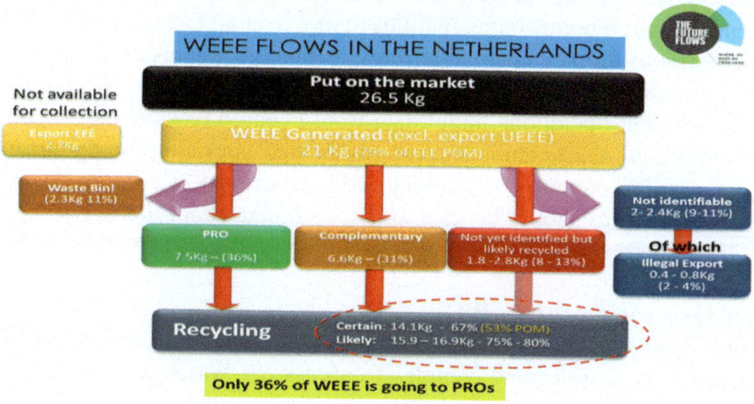

Fig. 3.5 Estimated WEEE flows in the Netherlands based according to Huisman et al. (2012)

Flows outside of the PROs can be significantly higher in countries with an active *Informal Sector*, e.g., in Africa or some Asian and South American Countries. The Informal Sector exists also in industrialized countries (e.g., Europe, USA).

E-Waste collection outside of PROs may include:

- E-Waste collected and managed by individuals, small companies (Informal Sector).
- Parts replaced during a repair or from remanufacturing operations are (usually) given to recyclers and properly treated, but not reported in the official collection data. This also includes products who cannot be repaired/remanufactured.
- E-Waste from replacing old installations is sold directly from *system integrators* to recyclers.
- The amounts of E-waste which are managed by commercial customers themselves (e.g., when products are replaced) without the involvement of Producers are usually not visible to the PRO/Clearing House.
- Municipalities are gaining financial benefits from managing valuable sections of E-Waste themselves. Germany is the only country where solid data about these amounts is available, up to 40% of the municipalities are managing at least one product category without involving Producers.
- Legally exported E-Waste is not included in the collection figures in many countries.
- E-Waste which is leaving the country without being declared/illegally exported.

Due to the lack of more sophisticated data especially in the early stages of E-Waste legislation, governments very often set collection goals based on E-Waste collected per inhabitant (kg/capita). Weight/Capita targets have been based on research about the amounts of E-Waste found in the generic waste stream or found orientation using similar goals established in other countries.

Some E-Waste legislation prohibits the collection of E-Waste by scrap collectors (e.g., Germany) or requires municipalities to turn over all E-Waste collected at their collection sites to PROs or individual Producers (e.g., Illinois).

In most countries, waste treatment licenses require recyclers to capture the in- and output of their operations. Input data from licensed recyclers are a convenient way to capture the waste amounts properly treated not requiring significant additional reporting. In conjunction with data of legal exports of used products as well as E-waste, countries should be able to get solid data about the E-Waste situation in their country.

E-Waste Generated/arising (WG)

The denominator in the formula to measure the Collection Rate (CR), E-Waste Generated (WG) is even more complex to access. "E-Waste Generated" is defined as all E-Waste discarded, independent if the waste is collected and treated properly or not. There is no place where to go to get exact data on the E-Waste Generated.

The amounts of E-Waste Generated are influenced primarily by the amounts of products sold in a country, the usage time (which can be different from country to country), the wealth of a society, the repair behavior, the potential for secondary/tertiary usage (e.g., to keep an old fridge for cooling beverages for garden parties) and many other socio-economic factors.

There have been a few attempts to include these factors into calculations for e-waste generated. For example, wealth of the population is influencing how long products are used. In countries with a lower GNP, the likelihood of the product replaced and discarded can be considered to be lower than in countries with higher GNP. Instead of throwing away the product, they might be handed over to people who are not yet in possession of such a device. Besides the national GNP (which somehow reflects the wealth of a population), cost of repair is an important variable to consider. The higher labor cost is in a country, the more expensive are repairs. If the price of a repair is reaching a certain threshold (e.g.,>42% of the price of the PC, 38% for a washing machine), customers will, most likely, discard the defective device (Hieronymi et al., 2020).

In industrialized countries, e.g., Germany, receiving a positive financial contribution from re-using products out of the waste stream is a challenge, even for social enterprises with low labor rates. The cost of, e.g., testing, repair, and warranty exceed in many cases the prices which could be achieved for these products (Hieronymi, 2022).

Academic Approach toward E-Waste Generated (*WG*)

As the real amount of *waste generated* is almost impossible to gather, several studies try to estimate the amount of waste generated involving various parameters. They are linked to a "discarded-based lifetime profile", using probability and distribution functions to calculate what amount has been discarded in a specific year (Magalini et al., 2016; Baldé et al., 2015). An in-depth review of the WEEE collection rate and a thoroughly analysis of statistical methods can be found at (Arduin et al. (2020).

There have been more research activities to determine more precisely the amount of E-Waste generated. Among them, the most comprehensive approach including an advanced, flexible, and multivariate Input-Output Analysis and several probability functions to generate more accurate data-sets as shown in Figs. 3.6 and 3.7 can be found in Wang et al., (2013) and Baldé et al., (2015).

$$E - Waste\ generatet\ (n) = \sum_{t=t^0}^{n} POM\ (t) * L^{(p)}\ (t, n)$$

Fig. 3.6 Formula to calculate the E-Waste generated by determining the amount POM and its assumed lifetime according to Wang et al. (2013), Baldè et al. (2015)

$$L^{(p)}(t, n) = \frac{a}{\beta^{\alpha}}(n - t)^{\alpha - 1} e^{-[(n-t)/\beta]^{\alpha}}$$

Fig. 3.7 Formula to calculate the lifetime of products using a Weibull function with a shape $\alpha(t)$ and a scale $\beta(t)$ Parameter

A full overview of the methodology and the explanation of the variables can be found in the guidelines on E-waste statistics from the global E-waste statistics partnership (Forti et al., 2018).

A good review of the challenges of any of these methods is discussed in the report "(W)EEE Mass balance and market structure in Belgium" (Wielenga & Huisman, 2013).

Simplified approach

The Collection Rate (CR) is a major measure of the success of any E-Waste legislation. The data required to set realistic goals and measure their achievement according to the complex formulas shown in the previous chapter is impossible in many countries.

Even with a simplified approach of calculating the return rate as shown in Fig. 3.8, there are huge challenges to get the data required.

$$CR = \frac{WCT}{WG}$$

Fig. 3.8 Basic Calculation of the Collection Rate (CR) by putting the Waste collected/treated (WCT) in relation to the waste generated (WG)

Many countries consider that the amount of waste generated (WG) is somehow related to the sales of products (S). While a generic connection between the amounts of product sold and the waste generated cannot be ignored, a replacement of WG by the sales of a certain year contains a few, but significant risks (Fig. 3.9).

$$CR_{(n)} = \frac{WCT_{(n)}}{S_{(n)}}$$

Fig. 3.9 In situations where no data on WEEE generated is available, WG is often replaced by the amount of products sold (S)

Stakeholders have realized that not all electronic and electrical products are used and therefore discarded in the country they have been sold in a first place. There are products leaving the country without being captured in national statistics (e.g., if an international company provides an IT infrastructure to their subsidiaries in other countries) or (legally) exported as 2nd hand products, e.g., from leasing/rental returns. Similar issues apply to direct sales across borders thru the Internet. Exports of IT products who are not registered in export statistics nor is the Producer aware that these products have left the market are e.g., Monitors of X-Ray machines, Notebooks built in tooling machines. In an interview, an expert from an IT company estimates that up to 10% of the sales in certain countries are exported (as a new product) without the knowledge of the Producer or the authorities.

The *net* number of products used in a country is called "Products Put on the Market" (POM) and reflects the local sales (S) minus legal product exports (E) (Figs. 3.10 and 3.11).

$$POM_{(n)} = S_{(n)} - E_{(n)}$$

Fig. 3.10 Basic calculation of POM by subtracting the amount of products exported from products sold

$$CR_{(n)} = \frac{WCT_{(n)}}{POM_{(n)}}$$

Fig. 3.11 Calculation of the Collection Rate using POM instead of the amount of products sold

However, even the number of products put on the market (Sales-Export) does not lead directly to the amounts of waste generated. Unlike packaging or beverage containers, electronic equipment is usually not discarded in the same year as it is purchased. In mature markets where a new product replaces an old one, sales figures may reflect somehow the potential for Waste Generated (WG). But products replaced by a new one may not end in the waste immediately. They may be kept in the household and may still be used occasionally for quite some time (e.g., a second fridge to cool beverages during summer) (Fig. 3.12).

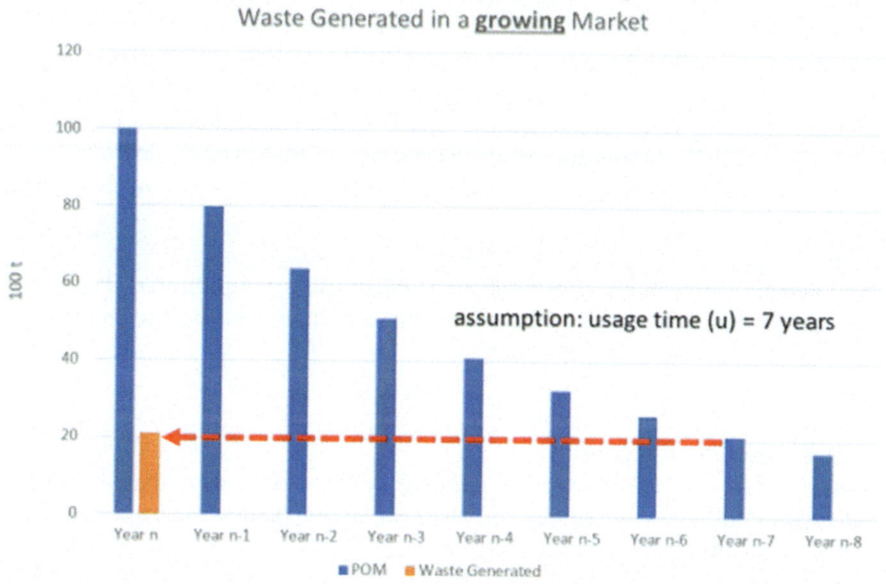

Fig. 3.12 In growing markets, Waste generated from EEE is significantly lower than the actual sales/POM

To reflect these impacts, some countries are using POM figures from previous years, reflecting an estimated usage time (u) for the electronic equipment (Fig. 3.13).

$$CR_{(n)} = \frac{WCT_{(n)}}{POM_{(n-u)}}$$

Fig. 3.13 To reflect the usage time in the calculation of a Collection Rate, the average usage time (u) is included in the formula

For example, the average usage time of a certain product group is 7 years. The target for the *Collection Rate* (CR) will be calculated by relating the *amounts collected and treated* (WCT) of the current year not to the POM of the current year (n) but to the amounts put on the market 7 years ago. This reflects that, in average, the products sold 7 years ago will be discarded this year. The reality is a little different as not all products will return in the same year. However, the impact of different usage times on the actual goal may not be significant as some products will be returned earlier, some in later years.

Usage time of EEE can be very different, reaching from a few years (low-quality small kitchen aid) to decades (high-quality fridges, power tools).

There are additional complexities which are very difficult to be captured. One of them is that new products take over the function of others. For example, watching movies, listening to music, and taking pictures has moved in many cases from what used to be called *Consumer Electronics* (CE) (e.g., radio, CD-players, stereo, cameras) to mobile devices like Smartphones and Tablet PCs, resulting in a significant drop in sales of traditional CE-devices. With old CE equipment still returned, but less sold, Collection Rates are increasing due to the dramatic decrease in the dominator (POM) of these products in addition to the impact of a reduced usage time.

Remark Calculating the Collection Rate using an adjusted POM (according to Fig. 3.13) is used widely as a basic approach to set more or less accurate Collection Targets.

3.1.4.6 Evidence Notes

In the early days of WEEE, the UK required Producers to ensure that all E-Waste collected by all actors, scrap dealers, charities, municipalities, and the Producer's own activities has been recycled.

At the end of the reporting period, Producers had to prove they had "*managed*" the recycling of a portion of the total amount of E-Waste collected, related to their market share. They either reported the tonnage recycled from their individual Take-Back programs and/or those "*their*" PRO collected and treated on their behalf. Waste companies who did not cooperate with a Producer or PRO could issue *Evidence Notes* for the waste they had treated. All actors could sell the certificates they did not need to those who did not meet their goals.

What started as a good idea, has turned into a near financial disaster. Stakeholders who were in possession of *Evidence Notes*, e.g., scrap dealers, kept them until the last minute, knowing that in a system where there is a demand for 100% of what is collected, any amount will find a customer who will pay almost any price (as long it is lower than the penalties applied) in the last minute of the reporting period. A new profession was created: "E-Waste Recycling Evidence Notes Brokers", accruing large sums of money without providing any additional value or contributing to a better environment.

To stop this situation, the UK abandoned the Evidence Note system. Instead, the UK government set fixed annual collection goals based on various factors (e.g., changes in the product mixture/design/technology/collection of previous years, import/export).

3.1.4.7 Convenience Target

Illinois has established a different approach to collection targets. The goal is that consumers can easily and convenient dispose their products. Producers are obliged to collect 100% to achieve this target. 100% of what is provided to the Producers by the citizens at *convenient* collection places (Consumer Electronics Recycling Act, 2017).

In this US-State, any district (county) is required to make a fundamental decision: either they cooperate with the EPR-System or manage E-Waste themselves.

As shown in Fig. 3.14, PROs must prove that they serve a minimum number of collection places according to population density, reaching from 1 in regions with less than 250 inhabitants/sqm to more than 25 locations where citizens can drop their electronic waste in more densely populated areas of this state.

Population density (people/square mile)	Minimum No. of Collection sites per County
< 250	1
251–500	2
501–750	3
751–1,000	4
1,000–5,000	5
Chicago-City	25

Fig. 3.14 Minimum number of collection sites in relation to population (Illinois)

Collection sites can be municipal waste collection places, shops, or sites set up by waste companies, or the PRO itself, they must meet minimum requirements (e.g., size).

There is no global formula to determine *Convenience*. The practical aspects (e.g., of the density of a collection network) depend on the individual situation in a certain country/region. Convenience is linked to the extra efforts people must take to discard their E-Waste properly. The more additional efforts required, the lower the convenience, and the more people are motivated to discard their E-Waste illegally. If they must drive multiple miles just to get rid of an old laptop, people may choose to discard it in a waste bin or, even worst, somewhere else. However, taking it to a place they often drive to anyway (e.g., their favorite shopping center), will be more convenient than bringing the retired device to a collection site closer affording an extra trip.

An alternative is *mobile collection stations*. An E-Waste collection truck stops, based on a fixed schedule at several places in a neighborhood (e.g., once every two months) and accepts the waste citizens bring (see Sect. 3.3.1.3).

The EU uses a split approach. On one side, they establish Collection Targets (specific to product categories) which are based on the relation between WCT and POM. On the other side, Producers are obliged to collect 100% of what is provided to them at municipal and other collection sites (e.g., shops).

3.1.5 Practical Approach for Setting Collection Targets

In practice, determining all variables of the formula presented in Figs. 3.6 and 3.7 requires data (e.g., age of the installed base, discarding practices) not always available (or very difficult to capture) even in countries where there is access to comprehensive statistics.

In practice, a few options have been proven to simplify the Target setting process for the waste collection.

▶ **Remark** Using these methods, even knowing they are not precise and have many weak areas, allow countries a quick start, focusing on setting up efficient operations first and refining the system after operations have been settled.

Convenience: 100% of what is returned

The target for the Producers in Illinois is to collect all products provided to them at collection sites. Similar targets have been established in the EU. The target is reached if no waste is left behind at collection sites. With the legislation providing guidelines on the density of the collection network a level playfield is created for Producers and their PROs.

The challenge is that there is no international practical measure for "convenience".

Convenience depends on the waste collection infrastructure, behavior of citizens, and factors like population density or the access to cars.

Start (s)low and accelerate

Another way to overcome the issues of establishing realistic collection goals is to start with low goals and increase them over time. A good example is South Africa's E-Waste legislation, enforced on Nov. 5th, 2021, with a collection goal in year one with 36.000t, growing with an annual rate of 30% to reach 103.000t in year 5 after the regulation has been put in place (Waste Act, Republic of South Africa).

▶ **Advice** Combining the 100%-Convenience Target with the "*Start (s)low and accelerate*" approach seems to be a good starting point for a new WEEE legislation. Enforcing a 100%-Convenience Target will ensure that no waste provided by customers will be left behind right from start. With more data gained during the operation of a WEEE-System in the first years, mid-/long-term collection goals can be developed, and realistic targets set.

3.1.6 Recycling Targets

Recycling Targets (*RR*) can be defined by comparing the amount of recycling materials resulting from treatment processes (*Output Recycling* (*OR*)) with the total amount of waste collected. More common is to compare the amount of materials recycled (*OR*) with the input into the recycling process (*IR*) (Fig. 3.15).

$$RR = \frac{OR}{IR}$$

$$RR = \frac{5.000\ t}{8.000\ t} = 62{,}5\%$$

Fig. 3.15 Principle of calculating Recycling Rates

As shown in the example above, calculating the Recycling Rate (RR) is simple, it can be done by relating the Output of recycled materials (OR) (5.000 t in Fig. 3.15) to the input into this process (IR) (8.000 t) of all registered recyclers. While the calculation is simple, it requires that recycling facilities exactly measure and provide reports on their input and output as well as those who are sent to incineration/disposal. In many countries, measuring the input/output and reporting it, e.g., on a regular basis to the authorities is an essential element of waste treatment licenses. In these cases, there is no additional effort required to calculate Recycling Rates.

▶ **Advice** When comparing recycling rates between different countries/jurisdictions, it must be noticed that Recycling Rates can be based on different definitions of *Recycling.* In many EU countries, using plastics as an alternative fuel (e.g., incineration with energy recovery in a cement factory) is considered recycling. If only *physical* recycling (w/o energy recovery) is included, current Recycling Rates of >90% for IT would drop significantly as recycled plastics from older products cannot be Re-used (e.g., due to brominated Flame Retardants).

3.1.7 Re-Use Targets

Any activity which is prolonging the usage time of products leads to waste minimization **is the preferred waste management strategy** (Article 4 of Directive 2008/98/EC).

▶ **Definitions** There are a few quality categories for products for re-use:

- **as is**
  Products are sold without a (professional) inspection, defects are not repaired, in most cases warranty is not available.

- **refurbished**
  Major functions of a product are checked, known defects are repaired. "*Cosmetical*" issues (e.g., scratches) are fixed. Some warranty is provided.

- **remanufactured**
  The entire product is tested, components with potential wear are replaced, Operating System is up to date. Full warranty is provided. The product's quality is equal to when it left the original production process.

Remanufacturing has been a common practice for high-end IT-Products for years. In recent years, the market for 2nd hand consumer products (e.g., Notebooks, Tablets, and Smartphones) increased significantly.

▶ **Remark** Re-Use in the context of E-Waste regulation is very often focusing on the re-use of products out of the waste stream. The entire 2nd hand market is usually not incorporated in re-use targets.

"Preparation (*of products from E-Waste*) for re-use" is considered a waste operation with, e.g., consequences to the classification of shipments according to the Basel Convention.

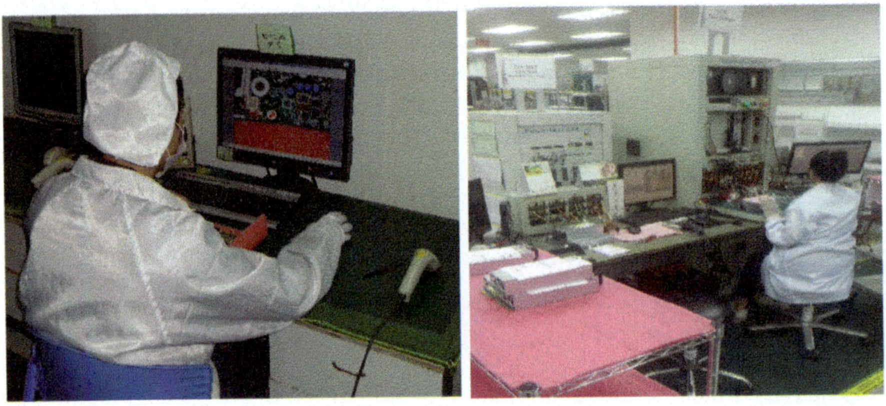

Fig. 3.16 Remanufacturing operations are using similar test equipment and processes as the original manufacturing lines

For the re-use of products from waste, the Re-Use Rate (RER) is defined as the relation between the amount of products re-used (PRU=32.5 t) and the amount of waste collected and treated (WCT=6.500 t) (Fig. 3.17).

$$ReR = \frac{32.5\ t}{6.500\ t} = 0.5\%$$

Fig. 3.17 Principle calculation of Re-Use Rates

Re-Use Rates compare the products picked out of E-Waste and sold, very often after being refurbished or remanufactured, as second-hand products. Re-Use targets do not include products sold on the *regular* second-hand market. The definition of re-use in connection with E-Waste legislation focuses on products who have been wasted before they are used, very often after repair/refurbishment again (Hieronymi, 2022).

It is a general belief that mobile IT and communication products (e.g., Notebooks, Smartphones) are only used for 2–3 years before they are discarded. The reality is different. IT products are used, by the same owner or after being handed over to a second/third user for more a long time. The products with re-use potential found in the waste at Australian and Danish municipalities are often quite old (7–8 years), show physical damage (e.g., cracks), and/or are incomplete (e.g., w/o external power supply) (Anthesis Group, 2015; Hieronymi, 2022).

Anthesis concluded that "Therefore, considering both a technical and economic reusability assessment, the percentage of products (*PC, Laptops*) that were re-usable is zero" at municipal waste collection sites they investigated in Denmark. The reason is that the majority of re-usable (IT/Communication) products have been moved to the second hand market before ending up in a municipal waste site.

Used IT products have, depending on their age, a significant re-sale value and are sold on the 2nd hand market or are handed over to family and friends. Only after multiple uses and several years, they might be discarded. While the 2nd hand market for laptops, etc. is mainly supplied by products sourced from large customers, there are pilots (e.g., in Shanghai, China) to test APM (Automated Purchasing Machines) for mobile devices from private users.

Maximizing re-use from waste requires significant investments in the infrastructure of Collection Sites. For example, instead of dropping a retired product in a container, products must be carefully stored in weather protected racks to avoid physical damage. It can be expected that depending on the product range of WEEE legislation (e.g., if large domestic appliances are included), an additional space of 40% will be required at collection sites. More staff may be needed to inspect products for potential Re-Use to direct citizens to either drop the products into a container are place it in a rack. It can be expected that the overall costs for E-Waste collection sites will rise by up to 2 times.

As the costs of remanufacturing/selling (and after sales support) devices from scrap are quite high, the demand for such a Re-Use goal often comes from social enterprises who have access to subsidized labor and therefore could handle old equipment with a very low re-sale value and high remanufacturing efforts. These products compete with the low-end of new products. According to a survey, customers may be willing to spent less than 20% of the price of a new one for a product originating from the waste stream. Brand new Smartphones could be found on the internet to sell for less than $85. Even charitable organizations may not be able to gain a positive financial contribution from handling units from waste (Hieronymi, 2022). The survey also revealed that many potential customers of secondhand products would refrain from buying one knowing that it has been "rescued" from the waste stream.

> **Remark** Re-using electronic products out of the waste stream will be only a small niche market. The age of the products in the waste, the cost of repair/refurbishment, warranty cost make these products unattractive, especially as they compete with the low price segment of new equipment which is available at very low prices (e.g., $85 for a new smartphone).

3.1.8 Summary (Targets)

Targets in context of E-Waste should be realistic, fair, and transparent. Data used for targets should be captured with a minimum of additional administrative burden. For Producers, a level playfield for Producers is essential.

These major elements should be considered when setting targets:

- Producers should be made responsible only for targets which they can influence.
- E-Waste properly treated (WCT) should recognize all waste streams who are directed into proper treatment, e.g., from PROs, Producers, Informal and formal waste collectors.
- In the first years, targets may be based on assumptions (e.g., using countries with a similar structure/culture/GNP as a reference) with the possibility to modify them based on the result and market development of the previous years.
- A "Convenience Target" (100% of what has been provided thru a (convenient) collection infrastructure) is a way to avoid complex calculations/data mining and minimizes the risk of E-Waste being dumped right from the beginning.
- Any Return Rate/Collection data should recognize Take-Back activities of individual Producers. They could gain, e.g., positive publicity by engaging with a charity project/setting up cooperatives within the Informal Sector, etc. For example, in conjunction with Take-Back events of large retailers in Germany (old for new discounts), sales raised by almost 300% that day.

- Recycling targets should reflect that some materials (plastics) contain substances which prevent their re-use (e.g., certain Flame Retardants).
- Realistic Re-Use targets are difficult to establish, as the number of products "arising from the waste" with a potential to be re-sold is very small, even in industrialized countries. It will be close to zero in developing countries as products that can be re-used will not appear in E-Waste and go directly into the 2nd hand market.
- To achieve collection targets, governments may request support from Producers. For example, some governments (France, recently Germany) ask the PROs to run communication programs to promote the proper disposal of E-Waste.
- If communication activities are requested to promote WEEE Collection, preference should be given to individual communication of Producers instead of common communication activities of all Producers or the PRO, allowing Producers to increase their individual sustainability profile. Individual communication will encourage Producers to include the E-Waste messages in their generic communication thus ensuring a wider spread and recognition. Individual communication requirements will also increase the acceptance within Producer's management.

3.2 Pricing Models

From a Producer's view, the financial model applied with Take-Back System is the major element to ensure efficiency of the system. Pricing models are an important element to ensure a level playing field and allow Producers to excel and gain a fair competitive advantage based on the attributes of their products. There are various combinations of pricing/financing models, but not all will be analyzed in this document. Ultimately, the **major impact of Take-Back legislation on Producers is the costs of managing their obligation.**

Across the globe, out of several, two basic pricing principles/models have evolved, Upfront Payments or Payments when Services are delivered

3.2.1 Upfront Payments

In a pricing model based on Upfront Payments, Producers are required to pay an amount per product they sell/put onto the market at the time when the products are sold/imported. These funds are used to finance the collection and treatment of the waste arising. Bills are sent prior to the collection/treatment and are due before any collection/treatment of a PRO takes place. There are various ways implemented for Upfront Payments.

3.2.1.1 Advanced Recycling Fee (ARF)

The most common form of Upfront Payments is Advanced Recycling Fees. PROs are charging a price for each product put on the market, Producers pay based on the number of products sold times the ARF as indicated in the example shown in Fig. 3.18.

		Put on Market	Producer Pays
Producer A *(30% market share)*		25,000 units	$20,400 *(25,000 units * $0.816)*
Producer B *(70% market share)*		58,333 units	$47,600 *(58,333 units * $0.816)*
Total units POM		83,333 units	
Estimated Cost of the PRO *(680 tons * $ 100 / t)*	$68,000		
ARF *($68,000 / 83,333 units)*	$0.816/unit		
Income of the PRO *(approx.)*	$68,000		

Fig. 3.18 Calculation of ARF within a PRO

ARFs can be different by product category and are set/controlled usually by the individual PRO.

They are usually calculated by including all costs and risks a PRO may experience. Major cost elements illustrated are in Fig. 3.19.

+ Estimated fixed cost of the PRO
+ Estimated PRO-Overhead
+ Estimated cost of Advertising & Promotion
+ Estimated Operational Cost ($/ton)
+ Estimated Collection (ton)
+ Accruals for uncertainties (e.g., Producers not paying their bills)

Total Cost (Budget)

→ divided by the estimated no of units POM = **ARF / Unit**

Fig. 3.19 Major elements used in calculating Advanced Recycling Fees (ARF)

PROs who operate with a pricing model based on upfront fees (Advanced Recycling Fees, ARF) need to predict various cost factors as they may change after the ARF is set and the Producers paid their bills (see Fig. 3.19).

It is a challenge to predict these factors especially in the first years. While the cost of collection and recycling can be secured in long-term contracts (e.g.,

12 months) with logistics and recycling companies, the number/weight of products Put on the Market could be based on asking their clients for their sales figures in the previous year. Estimating the amounts collected can be a huge challenge, especially as the collection rates differ significantly between product groups.

> **Remark** As PROs are not supposed to make losses, they are usually very conservative when estimating the major costs elements (e.g., higher return flows, lower number of products sold).
>
> As a result, most ARFs are calculated including a *safety margin* and are higher than the actual waste management costs.
>
> This lead in some PROs with an ARF pricing model significant more income. As a consequence, large reserves/accruals have been established.

In the Netherlands, in the first years of operation, accruals of up to €300 million were built up in the PRO from the surplus of the ARF charged to Producers. The remaining accruals of the PRO have been later confiscated by the Dutch government to be used in their generic budget.

Funds from ARFs for fridges, accrued by the Industry Association for Domestic Appliances in Austria prior to a WEEE legislation, have also been confiscated and shared for the use in their generic budgets by the Ministry of Justice and the Ministry of Environment.

Some government restricted the reserves which PROs could establish to reserves destined to cover the operational cost. They also establish a limit for these accruals to a maximum of 6 month of the PROs operational cost. The reserves will ensure that, in case of financial difficulties, a PRO can continue with their operations, e.g., to collect and recycle E-Waste for at least 6 months. These reserves will ensure that Producers can establish a new solution or move to another PRO in case more than one PRO exists.

3.2.1.2 Advanced Recycling Fees and Pricing Categories

Any pricing model should avoid advantages of one Producer or disadvantages of others. The goal is to establish a level playfield where prices are set to reflect the real cost of specific products.

The pricing model based on Advanced Recycling Fees bears the risk for cross subsidization when the same ARF is charged for products which cause different costs (e.g., collection and recycling cost).

Figure 3.20 is documenting a (fictive) situation where two Producers put each 25.000 devices on the Market. The products are similar but carry a different weight. The PRO assumed a Collection Rate of 60% for the products of both Producers, resulting in 105 t to be processed. Distributing the total Collection and Recycling Cost of $10,500 to both Producers based on the number of devices POM, an ARF of $0.21 was charged. This results in payments of both Producers of $5.250 each. In this example, *Producer B* paid $5.250, despite the fact that his products caused cost of $4,500 due to their lower weight.

	Units POM	Advanced Recycling Fee ($10,500 / 50,000)	ARF Payment
Producer A *(4kg/unit)*	25,000	$0,21	$5.250
Producer B *(3 kg/unit)*	25,000	$0,21	$5.250

Fig. 3.20 Advanced recycling fee for two Producers selling a product within the same ARF-category but with different weight/unit

Due to the differences in weight of the devices, the PRO has to manage 60 t of E-waste on behalf of Producer A, resulting in cost of $6000 while the cost for Producer B has been only $4,500. In this example, Producer B is subsidizing the collection and treatement cost of Producer a by $750 (see Fig. 3.21).

Collection Rate for both Producers = 60% of POM	Collection (Units)	Collection (weight)	Collection & Recycling Cost ($100/t)	ARF Payment	Cross Subsidization
Producer A (4kg/unit)	15,000	60 t	$6,000	$5.250	- $ 750
Producer B (3 kg/unit)	15,000	45 t	$4,500	$5.250	+ $750

Fig. 3.21 Specific cost caused by the different weight of individual products cause leads to cross subsidization

This is a simplified example on how the definitions within ARF pricing groups may impact the competitive playfield of Producers.

A solution often used in practice is to overcome/minimize cross substitution is to establish multiple ARFs for similar products but with a different weight. (In the example a different one for printers of 3 kg and those of 4 kg). However, more ARF pricing categories increase the administrative burden within a PRO as well as inside a Producer. The challenge is to find the right balance between the accuracy of ARF-Categories and the increased administrative burden who leads to higher cost.

More details about the impact Pricing/Product Categories may have on a level playfield can be found in Sect. 4.2.

3.2.1.3 Visible Fee (VF)

Visible Fees (VF) are ARFs who are made visible at point of sales to the customers. They were supposed to raise the awareness of customers for End-of-life management (EOL) of electronics and have been requested by some NGOs and governments, but also by some Producers in the early days of the discussions on WEEE in Europe.

Some countries requested the VF to be printed as a separate line item on the invoice, while others asked it to be shown on price labels and advertisements. In some countries, the visibility of fees is applied to all products, in others, it has been for commercial products only (e.g., France) or only for certain product categories (e.g., The Netherlands).

While the NGO's aim has been primarily on raising the awareness, the intention of some Producers has been to charge customers directly with the recycling cost and keep it out of the competition-battlefield of the products.

The attempt to exclude the Take-Back cost from the pricing of the product and therefore out of the responsibility of the Producers has been very intensive in the early days of the WEEE legislation when Producers expected waste management cost which could have severely impacted their bottom line.

There have been guesses in the early 2000s that the WEEE Management cost of players in the IT segment could raise to hundreds of million $/year for the EU only, indeed a severe impact on the profitability of Producers. Based on improved recycling technology, realistic data about collection rates, improved product design, and new regulation (e.g., ROHS) but also the increase in prices for raw materials and therefore revenues from recycled materials, todays cost are significantly lower (e.g., less than $0.5 for a low-end printer).

The following chapter illustrates the thoughts behind why some industries wanted to use the *Visible Fee* to move the cost of waste management away from the Producer.

Their thought has been based on a marketing expression:

"Psychological Pricing" or "Price Ending" (see *Wikipedia*) which is, simplified, the price the customer uses to compare products. If the price is $99,99, customers tent to consider it as significantly lower as a price of $100,10. Many Producers were afraid that the cost for WEEE Management would move the price of their products beyond the "Psychological Price Mark", forcing them to lower their profit margin to keep it.

Adding a fee at the cashier would allow them to keep the current price points (in the example at $99.99) as customers, as indicated in surveys, would ignore the additional price elements put on top of the product price. It is worthwhile to note that the discussion about the Visisble Fee started in countries with single PROs, where the same fee has been applied to products of different brands and therefore kept out of any competition between them.

► **Remark** Internalizing the waste management cost into the cost portfolio of the Producer is the essential element of Extended Producer Responsibility which the VF contradicts. In EPR, the individual Producer should be motivated to improve the design of a product which will lead to lower recycling cost and therefore its individual waste management cost.

Adding the waste management cost to the product as a Visible Fee, common to all producs of the same kind, will not provide any incentive to improve the design of products, especially in monopolistic structures of Take-Back Systems based on Collective Producer Responsibility where all Producers charge the same fee to their customers.

While many countries in the EU had been very interested in the visibility of the ARF in the early days of WEEE, most of them abandoned the request for a VF or allowed workarounds and/or do not enforce the *visibility* anymore. Producers noted that establishing a VF would require costly changes to their IT-Systems (e.g., invoicing) and regular maintenance in case product attributes change (e.g., weight, size) and the product falls into another VF-Category.

Waste Systems who kept the VF allow workarounds to simply the administration cost for Producers. Just a note that the "*recycling fees are included*" on the price label or invoice is the current practice, e.g., in Belgium. In Switzerland, Producers can refer to a list with the different ARFs published on the webpage of the Swiss PRO (SWICO).

According to the EU-Directive 2005/29/EG, published prices (e.g., in advertisements, brochures, webpages) must include all taxes and fees. Therefore, the goal of some industries to keep the E-Waste cost outside of *Psychological Pricing* is not possible as the final price has to be published anyway.

3.2.1.4 Tax

There are countries where waste management is considered a duty of the government (e.g., China, Taiwan). In these countries, Producers pay a tax to the government for products Put on the Market. The tax can be based on a percentage of the product price or a fixed amount per product sold. Taxes may be set differently for different product groups. What these taxes are put toward, is usually not transparent; taxes can be used for all types of government spending. Tax-based pricing models carry a few advantages (easy to administrate, no need for expertise in waste management within a Producer).

3.2.2 Payment When Services Are Delivered (Amounts Collected/Treated (ACT)

The alternative to Upfront Fees (Advanced Recycling Fees) is the pricing model the pricing model used in many WEEE-Systems:charging the Producer after the PRO fulfilled the Producer's obligation to collect and treat the E-waste he is responsible for. This pricing model can be called *Amounts Collected/Treated (ACT)*.

In a pricing model based on ACT, a certain amount of waste is assigned to a Producer who is responsible to collect and treat the waste according to the standards required.

In most countries using the ACT Pricing model, the amounts assigned to Producers reflect their market share.

The principles of the ACT model are documented in Fig. 3.22. The total amount of E-Waste collected will be distributed based on the market share of the two companies active on the market. In the example, with 30% market share, Producer A is responsible for 540 t while the PRO will manage 1,260 t for Producer B based on his market share of 70%. With $100/t, Producer A pays $54,000 to the PRO while Producer B is charged $126,000.

Collection & Recycling $100/t	POM (t)	Market Share	Obligation	Billed to Producers
Producer A	900	30%	540 t (30% of 1,800 t)	$54,000 (540 t * $100/t)
Producer B	2,100	70%	1,260 t (70% of 1,800 t)	$126,000 (1,260 t * $100/t)
Collection Rate: 60%			1,800 t	

Fig. 3.22 Simplified example on how the obligation of Producers are calculated in an ACT-based pricing system

In case of a higher Collection Rate (in the example 70% instead of 60%), Producer A's responsibility will increase to 630 t resulting in payments of $63,000. Producer B, based on his responsibility for 1,470 t, is supposed to pay $147,000. Pricing systems based on ACT are not causing any financial risk to a PRO caused by potential changes in the Collection Rate or POM as Producers pay for the obligation when the services are delivered, and the amounts put on the market and those collected are known (Fig. 3.23).

Collection & Recycling $100/t	POM (units)	Collection (t) (3 kg/unit)	Obligation	Billed to Producers
Producer A	300,000		540 t	$54,000
Producer B	700,000		1,260 t	$126,000
Collection Rate: 60%			1,800 t	
PRO Income				$180,000
PRO Cost				$180,000 (1,800 t * $100/t)
Profit / Loss				$0

Collection & Recycling $100/t	POM (units)	Collection (t) (3 kg/unit)	Obligation	Billed to Producers
Producer A	300,000		630 t	$63,000
Producer B	700,000		1,470 t	$147,000
Collection Rate: 70%			2,100 t	
PRO Income				$201,000
PRO Cost				$201,000 (2,100 t * $100/t)
Profit / Loss				$0

Fig. 3.23 The example shows that there is no impact on the financials of a PRO caused by increased collection in an ACT-based pricing system

In ARF-based pricing systems, fees are charged *in advance*. If the cost of the PRO increases in the same period, e.g., due to more E-waste collected than expected, adjustments cannot be made immediately. To avoid losses, PROs with

ARF pricing tend to include a safety margin in the calculation of ARFs to avoid losses as shown in Fig. 3.24.

Collection & Recycling $100/t	POM (units)	ARF / unit	Obligation	Billed to Producers [$]
Producer A	300,000	$=0.18		54,000
Producer B	700,000	$=0.18		126,000
Collection Rate: 60%			1,800 t	
PRO Income				180,000
PRO Cost				180,000 (1,800 t * $100/t)
Profit / Loss				$0

Collection & Recycling $100/t	POM (units)	ARF / unit	Obligation	Billed to Producers [$]
Producer A	300,000	$=0.18		54,000
Producer B	700,000	$=0.18		126,000
Collection Rate: 70%			2,100 t	
PRO Income				180,000
PRO Cost				210,000 (2,100 t * $100/t)
Profit / Loss				– $30,000

Fig. 3.24 Impact of increases in the cost of PROs, e.g., caused by higher collection than expected in an ARF-based pricing model

► **Remark** Pricing based on Advanced Recycling Fees provides a higher financial risk to a PRO then using a pricing model based on *Amounts Collected/Treated* (ACT).

3.2.3 Hybrid Pricing Models

There is a very interesting financial model proposed for TVs and Monitors in some US-States. Producers pay their PRO based on the amount of waste collected and treated allocated to them following the *Amount Collected/Treated* (ACT) pricing model.

In parallel, individual fees are charged at point of sales and transferred to Producers to cover their recycling cost. Depending on how good a Producer manages its obligation, there could be, in theory, a breakeven or even a surplus be generated. On the decision of the individual Producer, Waste Management Fees could be lowered or even set to $0, depending on the local marketing strategy.

The "Hybrid Pricing Model" allows to gain the best and avoids the negative elements of the ARF and ACT pricing models.

It allows Producers to gain competitive advantages by DFE and smart management of Take-Back costs. Other than in the ARF model, fees are not flowing to the PRO but to the Producer directly and can be used to (partially) fund their DFE and Take-Back costs. Reserves do not have to be established (unless requested by the internal controllership of a Producer).

3.2.4 Comparison of Pricing Models

There is no generic rule on which of the pricing model is to be preferred. A choice depends very much on if a Producer has the skills to manage partners who provide the operation of his WEE-obligation, the magnitude of his Take-Back costs, and if these costs are managed within a Producer on a local level or an international basis. Usually, Producers prefer a flexible solution and select the pricing model which fits best to the local situation. Therefore, pricing models should not be determined by the legislation to allow Producers to choose the PRO who's pricing model fits best to the Producer's needs.

Any legislation should allow that individual collection and treatment operations of Producers are recognized against their obligation. Collecting E-Waste, e.g., in conjunction with sales events will not only increase the sensibility of consumers of this topic but also allow the Producer to focus the collection on his own products and act, at least in a small scale, as within the IPR principle.

In the case of an ARF-based pricing model, attention should be paid to set up fair ARFs, avoiding cross-subsidies with competitor's products or products from different industry sectors.

The more ARF categories are established, the more product attributes who lead to changes in the cost envelope can be recognized in the ARF setting. However, the more different ARF categories are set, the higher is the burden to manage them within a PRO as well as the administrative cost within a Producer. For example, Producers need to check on a product-by-product basis what category to apply for new products (in the IT-Industry products are invented on a monthly basis) and if any changes within existing products lead to a different categorization (e.g., changes in weight).

There should be a balance between the number of different ARFs for individual categories to recognize different transportation/recycling costs/weight on one side and administrational effort to manage many ARFs.

For example, SWICO, the Swiss PRO, has set up 10 different ARFs for monitors based on size (and therefore indirectly weight). Producers are required to evaluate every new monitor manually (e.g., checking the product data sheet) for the screen size to set the appropriate ARF. To the Belgian PRO, RECUPEL, charges the same ARF for all monitors, independent of cost relevant parameters like size and weight. Producers in Belgium will have less administrative efforts but experience cross over subsidies between Producers with different sizes of monitors.

Advanced Recycling Fees are—for Producers—easy to administer, depending on the number of product categories and if the assignment of products to the pricing categories can be automated. Joining a PRO with an ARF pricing model does not require any specific knowledge of the Producer for waste services (e.g., collection, recycling). The selection of the PRO (if possible) can be handled by a procurement department without any skills/knowledge in the waste market.

However, there are also some disadvantages of ARF pricing models:

- Difficult to reward individual collection of Producers.
- In case of legislation, who moves the obligation to manage waste from the Producer to the PRO, the PRO must cover the cost of managing waste of Producers who fail to pay their bills.
- Precautional reserves for uncertainties in forecasts of cost (e.g., the tonnage collected and the number of products sold) need to be established and will lead to higher prices and accruals which may rise the "appetite" from governments.
- There is a risk of financial instability of the PRO in case of unexpected developments in sales figures/Collection Rates leading to a loss of the prepaid fees.

In case the ARF is established with the concept of a Visible Fee, Producers need to adjust their IT-Systems to include the VF. This can be, depending on local rules (e.g., different VAT on the fees and the product itself) and the fees structure request sever changes. There are reports that changing the internal invoicing systems to match, e.g., the Hungarian packaging waste requirements required multi-million $ investments of a single Producer.

If a country chooses a tax-based pricing model, these challenges can be observed:

- Slow to adopt changes (e.g., recycling costs decrease).
- Funds raised may not be used for recycling/sustainability.
- No competition.
- Own collection activities of Producers are not rewarded.
- Tax-setting not transparent.

A pricing model, where the Producers pay for the management of the amount of waste they are assigned to (*Amounts Collected/Treated* (ACT)), carries a significantly smaller risk of cross subsidiaries than with ARFs. In this model, there is no accountability of Producers for those who do not fulfill their obligations. For a PRO who is using the ACT Pricing model, financials are much easier to manage. In case they align the pricing periods between their suppliers for the collection and treatment with their clients, almost no accruals are required. Changes in collection will automatically be transferred to the Producers who pay after the service is delivered, not upfront as in an ARF-based pricing model.

Administration for Producers is simpler with ACT-based pricing as Producers must report the weight of the products put on to the market based on collection categories. There is no need to report, e.g., for different weight groups within a product/collection category. On the downside, the Producer, in an ACT-based pricing model carries the risk of changes of the amounts collected and in case treatment/logistics cost change.

> **Conclusion** *Advanced Recycling Fees* (ARF) require, especially in the first years of a WEEE legislation, many estimations as solid data are not yet available (e.g., E-Waste Generated, future collection and treatment cost, composition of the waste), estimations which are quite difficult to set up. As PROs are not supposed to make a loss, they tend to include a safety margin into their calculations. With gaining experience over time, these safety margins could be minimized.
>
> PROs using a pricing model based on *Amounts Collected/Treated* (ACT) do not require fundamental estimations for the major cost elements. The amounts collected are charged out based on the actual collection. Therefore, systems based on an ACT pricing model do not require safety margins in their calculations. ACT-based pricing minimizes the financial risk to a PRO significantly.
>
> TAX-based pricing models should be avoided unless they provide transparency on how the funds are used.

3.2.5 Economic Rewards for "*Design for Recycling*" (Modulation)

Rewarding *Design for Recycling* (DFR) is an essential element of the *Extended Producer Responsibility* principle.

From an academic perspective, *Individual Producer Responsibility* (IPR) (manufacturers taking only their own products back) is the appropriate thought model but implementing it in practice creates many challenges. There is currently no practical means to implement IPR as the efforts to separate waste by brand outpaces the benefits (e.g., Japan where the cost of collecting/recycling a PC is>$20).

The EU is currently debating a strategy (developed in France) to reward DFR, even if recycling costs advantages cannot (or only to a very small extend) be realized in a Take-Back System based on Collective Responsibility. This strategy is called *Modulation.* Hazardous components whose removal has been the cause of higher recycling cost have been significantly reduced in all modern EEE (enforced by, e.g., RoHS, EPEAT). Consequently, differences in recycling costs of products designed with and without DFR are usually very small and it is very difficult to quantify them in an operational environment where the waste collected contains both, products designed with and without recycling in mind. The option to process only products where DFR has been applied would require sorting and separate treatment. It may also require developing new recycling processes to gain from the changes made in the design of products. If at least in early stages of a Take-Back System, the majority of the waste collected will consist out of old products where DFR is not applied, batches of products with DFR will be small. Together with the sorting required, the cost of separate processing will eliminate potential cost advantages for products designed for optimized recycling.

In the absence of significant, proofed collection and recycling cost savings, other criteria may be applied. However, it is very difficult to apply an economic value to these criteria (e.g., repairability).

In these cases, *Modulation* is more a political instrument to encourage certain design directions than based on real economic effects according to the principles of Extended Producer Responsibility. Biggest challenge for **Modulation** will be to harmonize these criteria among countries as almost all electronic products are produced for global markets and taking care for specific country requirements are almost impossible.

3.2.5.1 Modulated Advanced Recycling Fees

In an ARF-based pricing model, ARFs could be lowered for products that are easier to recycle and therefore cause lower recycling cost. Those who fail to follow DFR guidelines could see their costs on the raise. Implementing modulation could be applied even if DFR and Non-DFR-products are collected and treated together, and potential recycling costs advantages cannot be formally recognized. Including additional environmental elements into the modification of the ARF (e.g., reparability) can change the nature of an *Advanced Recycling Fee* into a kind of tax, providing a bonus-malus model to reward generic environmental features of products (Fig. 3.25).

Modulated ARF	Product A	Product B	Product C
"Environmental Score"	+5	−4	-9
Original ARF	$0.5/unit	$0.5/unit	$0.5/unit
DFR "Malus/Bonus"	−5%	+4%	+9%
Modulated Obligation	$0.48/unit	$0.52/unit	$0.55/unit

Fig. 3.25 Table illustrating the principle of "Modulated Fees". The "Environmental Score" is supposed to reflect several product attributes impacting the environment (e.g., Repairability)

While the principle of *Modulated Fees* may be considered an instrument to drive Producers toward DFR, it carries some disadvantages:

– The composition of the waste which PROs are managing will contain a mixture of products with high and low environmental scores. It is a valid assumption that all PROs operating in a country will receive a similar mix of waste and therefore are charged with the same treatment cost.

- PROs will therefore be eager to get clients whose products have a low environmental score as they can charge higher fees. With the same operational cost, their income will increase compared to PROs whose clients put product with high environmental scores onto the market.
- For establishing a level playfield, funds need to be transferred between PROs to eliminate a negative impact on their finances due to modulated fees.
- Modulated Fees prefer (almost exclusively) pricing models based on ARF.

3.2.5.2 Modulated Collection Obligation

While *Modulation* is usually coupled with an ARF-based financial system, it is not limited to this financial model. It can also be implemented in Take-Back System based on Payment by *Amounts Collected/Treated* (ACT). Instead of modulating the fee of a product (=> ARF), the collection obligation of Producers is *modulated*. Those Producers whose products receive a lower environmental score must collect more E-Waste than their market share would lead to. On the other hand, to compensate for their DFR investments, the collection and treatment obligation of Producers with more sustainable products will be reduced as shown in Fig. 3.26.

Modulated Collection Obligation	Producer A	Producer B	Producer C	Total Obligation
Environmental Score	+5	-4	-9	
Original Obligation	6,000 t	3,000 t	2,000 t	11,000 t
DFR "Malus/Bonus"	-5%	+4%	+9%	
Modulated Obligation	5,700 t	3,120 t	2,180 t	11,000 t

Fig. 3.26 Illustration of the "Modulated Collection Obligation" principle

Modulated Collection Obligation:

- allows both pricing models, *Advanced Recycling Fees* (ARF) as well as *Amounts Collected/Treated* (ACT).
- is not disturbing competition. PROs who are serving Producers with high environmental scores will manage lower amounts of waste, hence have lower treatment cost. PROS with many clients with low scores will process more e-waste and therefore have higher cost which is compensated as they can charge the extra amounts to these clients. Modulated Collection does not interrupt the level playfield between PRO or preferences of PROs to acquire Producers whose products carry low environmental scores.

3.3 Collection

Convenience is an important element to motivate customers to move their retired products to proper collection channels and therefore ensure appropriate Collection Rates. Even customers who care about protecting the environment may put an electronic product in the waste bin if the efforts for proper disposal are significant.

On the other hand, convenience has its price. The denser a collection network is, the higher the collection costs.

Convenience cannot be defined on a universal basis. It is determined by many local/cultural/individual elements which differ by country and region (e.g., existing collection infrastructure, access to cars, etc.).

Basically, two models for the collection of E-Waste have evolved,

- either the E-Waste is collected from the customer's household (collection from customers) e.g.,:
  - On Demand Pick-Up at Customer Site.
  - Curb-Side Collection.
  - "Milkman-Tour".
  - Waste Pickers (Informal Sector).
- or the E-Waste is brought by the customer to a collection location (Bring In), e.g.,:
  - Collection Sites.
  - Dealers/Shops.
  - Take-Back Events.

3.3.1 Collection from Customers

3.3.1.1 On Demand Pick-Up at Customer Site

The most convenient collection method for retired electric and electronic equipment is "On Demand Pick-up" from the customer's home. Customers inform the Producer (or his PRO) by web or phone about the products they would like to dispose. A collection service picks up the products and provides them (e.g., thru consolidation warehouses) to the treatment facility, the Producer/PRO is cooperating with. Producers pay for the services on an item-by-item basis. Such a system is currently established for PCs and monitors in Japan, the average cost is more than $20-per unit returned. A similar system has been established for large domestic appliances in Korea, with the exception that the collection is not done with a courier service. The logistics provider is arranging tours to collect from consumers of a neighborhood not only for electronics but for all WEEE.

3.3.1.2 Curb-Side Collection

In a Curb-Side Collection system, citizens place their E-Waste on the sidewalk in front of their property (=>Curbside) where it is collected on a regular basis. The dates of collection are either communicated or there are fixed dates (e.g., every 1st Monday of a quarter).

3.3.1.3 Collection System "Milkman-Tour"

Collection by *Milkman-Tour* requires slightly more efforts by the citizen then than Curb-Side Collection. Waste is not picked up at each doorstep. Citizens can dispose E-Waste at waste collection trucks who stop at central points in neighborhoods (e.g., local shopping center, sports ground, parking lot) on a regular basis. *Milkman-Tours* are very often used to collect hazardous waste within municipalities (Fig. 3.27).

Fig. 3.27 The picture is featuring a truck collecting hazardous waste from households by Milkman-Tour in Germany (picture courtesy of Remondis, Germany)

3.3.2 Bring In

3.3.2.1 Collection Sites

Collection sites are the most common way to collect E-Waste in countries with an organized waste management infrastructure. E-Waste is brought by the customer to public collection sites who are accepting also other wastes. Collection sites are very often set up by municipalities. They are, in many cases located outside of city centers in industrial zones, therefore requiring access to cars to get the E-Waste there.

Collections sites for E-Waste could also be set up by PROs or private enterprises (e.g., scrap yards). Several countries in Asia include criteria for collection sites (e.g., workers safety, fire safety, pollution prevention, security) in their local WEEE/Recycling legislation (e.g., Standard AS/NZ 5377).

3.3.2.2 Collection at Retail (Shops)

To increase the density of a collection network and therefore the convenience level for citizens, the latest version of the EU-WEEE-Directive requires retailers to accept E-Waste from clients. This is limited to retailers with a shopfloor of 400 sqm and above (approx. 4,280 sqft). The retailers have to accept E-Waste from the type of products they sell, but not necessarily related if the product has been sold by this retailer or not.

3.3.2.3 Take-Back Events

Different from other forms of E-Waste collection, "Take-Back" events are not scheduled on a regular basis. They can be organized by municipalities, Producers, or PROs. In a Take-Back Event, customers bring the retired equipment and are rewarded for example with a discount on the purchase of a new product or participation in a contest/raffle. Take-Back Events can be an instrument to get access to waste from selected products, e.g., equipment of a specific Producer.

Various rewards have been offered and tested to be successful:

- Discount on the purchase of a new product
- Competition (e.g., sport clubs who provide the largest amount of waste, the oldest product)
- Ruffle/Draw (price, e.g., a new printer)
- Cooperation with charity organizations (e.g., functional equipment donated to them)

Experience in several events showed that, when products are checked for re-use and provided to a charity organization, the returned products are usually younger and of better quality.

3.3.3 Density of a Collection Network

The denser a collection network is, the more convenient it is for citizens to use. In countries with an already established waste collection infrastructure (e.g., Europe, North America) it has been a common belief that the existing network of collection sites provides sufficient convenience for citizens. However, as the amount of WEEE disposed in household waste (especially small items) did not decrease significantly over time, the second version of the WEEE-Directive extended the collection network by including retail shops. This extended the waste network more than threefold. To what extent this has an impact on the amount of WEEE disposed in waste bins (household waste) is not known yet.

In most jurisdictions, the density of the collection network is subject to negotiations between PROs, Producers, and the authorities—considering population density, existing waste collection infrastructure, access to cars, and habits of citizens. There is no magic formula which fits everywhere.

3.3.4 What Type of Collection to Choose

▶ Rule of thumb: the more convenient a collection infrastructure for waste is, the more is collected, but the higher is the collection costs.

A Take-Back System based on individual pick-up at a customer side is the most convenient, but also the most expensive system due to the high logistics costs. With more than $20/PC or Monitor collected, the Japanese TBS is the costliest system around the globe (on a per item/collected basis).

Curb-Side Collection requires special trucks, routes, and personnel. Citizens use to drop not only E-Waste at Curbsides but all kinds of waste. Therefore, multiple trucks/trailers and personal to sort the waste when loading are required with Curb-Side Collection. Curbside Collection of E-Waste requires also sorting the E-Waste into the Collection categories, increasing the complexity of the sorting process and the number of vehicles required. Experiences in many countries show that waste provided for collection at the curbsides is often subject to search and cannibalization by all kinds of individuals, leaving piles of waste distributed over sidewalks and streets (and therefore becoming a danger to both pedestrians and cars) (Fig. 3.28). Costs for Curb-Side collection of E-Waste are at least three times as high as the costs of a Collection Site (located outside of a city center) (on a per ton basis).

Fig. 3.28 “Raided” pile of waste with “Curb-Side Collection”

Costs of a *Milkman-Tour*-based collection service are slightly lower than Curb-Side collection. It still requires special trucks and trained personnel but involves less mileage as only certain places in a neighborhood are visited. In addition, the risk of *raided waste piles* is low as citizens drop the E-Waste into the truck while the staff is present, and the waste is not kept unattended at the curb-side. There are no concrete cost data available for the *Milkman-Tour* approach.

Collection systems based on Waste Collection Sites are the most common way to collect E-Waste in industrialized countries, where most communities have these sites already established. At most collection sites, citizens are requested to place their E-Waste into standard containers or waste skips separated by type of equipment according to local rules.

The minimum additional space required at collection sites for the collection of the six E-Waste Collection categories (containers plus space to load/unload) in the EU is about 3000 m^2 (approx. 32,000 sq ft).

Rental prices for space and containers, investments in buildings, machinery, and costs of staffing are major cost elements of Waste Collection sites. Running an E-Waste Collection Site may cost approx. >$25,000 (EU countries) per month mainly depending on the number of staff involved and labor costs, but also where it is located (rental cost for space). Including generic infrastructure costs like access control, investments in safety/security (fencing) and staff related buildings are increasing the costs significantly.

In some countries PROs pick up E-Waste free of charge from municipal collection sites, in others they must provide the infrastructure (e.g., containers) to collection sites, in a few countries (e.g., France, Austria), the municipalities raise a fee for every container handed over to a PRO. In Switzerland, the PROs had to organize their own network of collection sites, partnering, e.g., with private companies and municipalities.

At retailers, depending on their location, establishing an E-Waste facility is more expensive, especially for shops in city-centers where space/shop-floor is limited. Some of the shops are using parts of their receiving deck, others use parking spaces. Some of them are offering free pick-up in case a new product is bought and transported to the customers address (e.g., large domestic appliances) as an alternative. Either way, establishing the capability to collect E-Waste at (or close to) a shop is not cheap.

The costs of a Take-Back Event depend to a large extent (excluding renting space) on the rewards provided. Theoretically, a discount on the sale of new products is quite costly. However, in conjunction with marketing activities, collection events can be profitable for the Producers involved. In addition, these events are contributing to a positive image of these Producers. It has been reported that sales, in conjunction with a discount, raised threefold at Take-Back Events of retailers.

3.3.5 Informal Sector

The "Informal Waste Management Sector" comprises individuals who are involved in the collection and treatment of waste. Individuals who are (based on a strict interpretation of the expression) neither employed nor hold a business/waste collection license. In connection with E-Waste, this expression is used to include all collectors who are not directly serving a Producer or a PRO, including very formal enterprises such as scrap dealers.

While the *Informal Waste Management Sector* can play a major role in collecting E-Waste, measures should be taken to ensure that the collected waste is delivered to facilities that apply proper recycling standards and focus on the protection of the employees from health hazards (Fig. 3.29). Financial incentives, e.g., payments to the collectors for untreated E-Waste which are higher than what they could gain from illegal treatment are one of the options to take.

Fig. 3.29 Especially in developing countries, collectors are involved in "below standard" treatment of E-Waste. Fotos taken in Ghana and India by Matthias Schluep

The Informal Sector is a major pillar of E-Waste management in countries like Brazil, Ghana, India, Nigeria, South Africa, Vietnam, India, and the Philippines (Li et al., 2013; Vaccari et al., 2019). In some countries, their services are integrated into Take-Back Systems (e.g., Brazil, India) or will be soon (Section 18 of the National Waste Management Act of South Africa (NEMWA)).

The Informal Sector is actively collecting from homes, curb site piles, repair shops, and small companies. Their service is very convenient for their clients as the waste pickers take mixtures of different kinds of waste (e.g., cardboard, furniture, plastics, E-Waste) waste whom they sort. While the collection service of the Informal Sector can help to increase the Collection Rate, it has to be ensured that they deliver the collected E-Waste to proper treatment facilities and do not capture the valuable materials by using improper processes (e.g., burning cables) themselves. Burning electrical/electronic components releases many toxic substances, namely, due to the presence of Chloride or Bromine (which are contained in plastics from pre-RoHS products) from the flame retardants of plastics, toxic Dioxins. Solving metals from PC-boards and other components from electronic equipment with, e.g., acids provide additional health hazards to workers and bystanders. These operations do also release heavy metals like lead which pollute the surface/ground water but also the soil neighborhood of these operations.

There have been several pilots (e.g., in Africa, India) to establish rewards for the collection services of the Informal Sector, rewards high enough to prevent illegal recycling practices. Education about the health risks of improper treatment can assist this intention.

Rewarding the collectors above what they could gain in improper treatment carries the risk that *middleman* stores the E-Waste and sell it to the highest bidder. Especially in countries who have set unrealistic collection targets, this behavior carries the risk the cost for Producers raises without any benefit to the environment of the people who collected the E-Waste.

> **Advice** While the *Informal Waste Management Sector* can play a major role in collecting E-Waste, measures should be taken to ensure that the collected waste is delivered to facilities that apply proper recycling standards and focus on the protection of the employees from health hazards. Financial incentives, e.g., payments of the collectors for untreated E-Waste which are higher than the result of illegal treatment are one of the options to take.

3.4 Clearing

Any Take-Back System based on Collective Producer Responsibility requires some form of administration (e.g., to register Producers) which, in the context of Take-Back legislation, is often called *Clearing* or *Control Tower*.

Tasks which Clearing can include:

- Registration of Producers and PROs.
- Document the amounts put on the market by each Producer.
- Calculation of the market share of Producers.
- Calculation of each Producer's Take-Back obligation.
- Insurance that Producers/their PROs have access to the amount of waste required to fullfill their obligations.
- Assignment of collection sites to PROs.
- Moderation the compensation between PROs.
- Control of the obligation achievement.
- Audit Producer's individual collection/treatment.
- Provide reports of non-compliant of Producers to authorities.
- Reporting achievements of various targets to authorities.

> **Remark** "Clearing" is required independently if a single or multiple PROs are operating in a country. While in a Take-Back System with a single PRO, Clearing can be performed within the PRO, a Clearing function independent of the PROs is required in a system based on multiple PROs to ensure a level playfield for them.

According to an overview provided by the OECD (2016), several governance functions could be allocated to stakeholders in Take-Back Systems (Fig. 3.30).

Governance Function	Govern-ment	PROs	Clearing House	Commercial Companies (e.g. logistics, recycling)	Munici-palities
Policy Formulation	*				(*)
Operations				*	
Registration			*		
Accreditation	*		*		
Of PROs					
Collection of Producers financial contribution		*			
Coordination			*		
Monitoring			*		
Reporting to Government			*		
Preparation of Enforcement			*		
Enforcement	*				(*)

Fig. 3.30 Allocation of governance tasks, based information provided by the OECD (2016)

3.4.1 Registration of Producers and PROs

In order to ensure Producers are fulfilling their Take-Back obligation, it's essential that all Producers are registered. To avoid high administration costs (relative to the duties of a Producer), some countries give exceptions to very small companies, e.g., if their imports are below a certain amount or value.

In this context, special attention should be paid to importers who disappear after a few months of operation and may start again under a new name and therefore avoid contributing to the collection and treatment of E-Waste.

► **Advice** A simple but efficient way to control that Producers are fulfilling their collection/treatment obligation is to ask them to state their registration status (e.g., a registration no.) on their invoices. Or request PROs/Clearing House to publish the list of registered Producers (e.g., Switzerland: Konventionsunterzeicher; Germany: EAR Herstellerverzeichnis).

Some countries require a registration certificate for customs clearance. This is also a very efficient way to ensure that all Producers are registered.

3.4.2 Calculating Producers' Obligations

Determining the actual obligation of a Producer is a major function of Clearing.

In a Take-Back Structure based on *Collective Producer Responsibility* (CPR), the market share of the individual Producer drives its obligation. Consequently, the sum of its client's obligations is determining the amount a PRO must collect and treat.

This model is, with small variations, implemented in the majority of WEEE legislation across the globe.

In a first step, the share of each Producer on the overall amounts Put On to the Market (POM) is determined, based on the amounts they reported. Figure 3.31 shows how the different market shares of Producers are calculated. For example, *Producer A* puts 2,500 t on the market, resulting in a POM Share of 25% of the total market of 10.000 t. For *Producer B* the share is 18%, based on 1,800 t he puts on the market. *Producer C*'s 1,600 t resulting in a market/POM share of 16%.

	Clients of PRO I				**Clients of PRO II**		
Producer	POM (t)	POM Share	Obligation (t)	Producer	POM (t)	POM Share	Obligation (t)
A	2,500 t	25%		B	1,800 t	18%	
D	1,300 t	13%		C	1,600 t	16%	
E	1,000 t	10%		G	500 t	5%	
F	700 t	7%		I	200 t	2%	
H	400 t	4%					
Total POM **10,000 t**							

Fig. 3.31 **Step 1, calculating each Producer's POM share**

In step two, the total collection obligation is calculated based on the actual collection or the collection obligation given by the authorities. The total collection obligation (in the example illustrated in Fig. 3.32) 6,254 t, is distributed to all Producers based on their market shares calculated in step 1. Based on this calculation, the obligation of *Producer A* is to pick up and treat 1,564 t while the obligation for *Producer B* is 1,126 t, respective 1,001 t for *Producer C*.In a third step, the collection obligation of the clients of a PRO is added up resulting in the tonnage of E-Waste each PRO has to collect and treat on behalf of its clients.

Total Collection (Obligation) 6254 t							
	Clients of PRO I				**Clients of PRO II**		
Producer	POM (t)	POM Share	Obligation (t)	Producer	POM (t)	POM Share	Obligation (t)
A	2,500 t	25%	**1,564 t**	B	1,800 t	18%	**1,126 t**
D	1,300 t	13%	**813 t**	C	1,600 t	16%	**1,001 t**
E	1,000 t	10%	**625 t**	G	500 t	5%	**313 t**
F	700 t	7%	**438 t**	I	200 t	2%	**125 t**
H	E 400 t	4%	**250 t**				

Fig. 3.32 Step 2, calculating each Producer's collection obligation

Total Collection (Obligation) 6254 t							
	Clients of PRO I				**Clients of PRO II**		
Producer	POM (t)	POM Market Share	Obligation (t)	Producer	POM (t)	POM Market Share	Obligation (t)
A	2,500 t	25%	1,564 t	B	1,800 t	18%	1,126 t
D	1,300 t	13%	813 t	C	1,600 t	16%	1,001 t
E	1,000 t	10%	625 t	G	500 t	5%	313 t
F	700 t	7%	438 t	I	200 t	2%	125 t
H	400 t	4%	250 t				
PRO I Collection Obligation			**3,690 t**	**PRO II** Collection Obligation			**2,564 t**
Total Collection 6,254 t							

Fig. 3.33 Step 3, calculating the obligation of a PRO by adding the obligation of its clients

To encourage Producers to set up and run individual take back initiatives, e.g., with providing an incentive when an old product is returned with the purchase of a new one, the volumes of an individual Producer Collection program should be subtracted from the amounts a PRO is required to collect for this Producer. At the same time, these amounts need to be included in the overall colletion amount. However, the individual collection activities should be monitored to avoid fraud.

3.4.3 Access to Waste

Many stakeholders have access to E-Waste: municipalities, scrap dealers, waste pickers (Informal Sector), PROs, sales channel partners, recyclers, large customers, Producers—just to name a few.

In countries with a monopolistic Take-Back Structure, all waste provided to Producers will be managed by one PRO.

In systems with more than one PRO, ensuring that each PRO has access to the amount of waste they are supposed to manage for their clients is essential.

Clearing functions can ensure that each PRO has access to the amount of waste they need to manage to fulfill the obligation of their clients.

Basically, there are three major strategies for the allocation of waste between Producers/PROs:

- Not Regulated (No allocation of waste/compensation).
- Allocation of Individual Containers.
- Allocation of Collection Sites (or regions).

3.4.3.1 Not Regulated/No Allocation Mechanism

There have been countries where the WEEE legislation/implementation rules did not foresee any of the Clearing functions; no central registration, no "access to waste rules", no reporting of total amounts to be collected, and no over-/under-collection reporting or compensation mechanism.

The only obligation of a Producer has been to become a member of a PRO.

Without clear obligations existing, fair playing PROs collected E-Waste based on estimates about the total amount of waste based the (also estimated) market share of their members.

Other PROs collected only a very small amount, enabling them to offer compliance at a very low price.

This led to an unfair situation in the early days of WEEE in, e.g., Spain, where setting and controlling targets is the responsibility of the 19 regions. Some regions did establish targets for PROs (and enforced them), while others did not.

After a few years and, e.g., legal interventions by the PROs acting fair, regional/central governments established clear collection/treatment responsibilities for PROs.

3.4.3.2 Allocation of Individual Containers

To avoid potential differences in transportation costs (e.g., between rural and urban areas), the German authorities (with strong support from the industry) developed a system based on the allocation of each individual container, independent of its location.

In practice, a Collection Site reports a full container to a central organization (*Elektro-Altgeräte-Register*—EAR) who assigns it, based on an algorithm that takes the market share and the collection history of a Producer into consideration, to a Producer. The pickup request is then routed to the Producer's PRO. This PRO organizes the pickup of the container, treatment of its content, and charges the Producer for its services according to their contract. The algorithm developed by EAR ensures a fair allocation of pickups from rural and urban areas across Germany for each Producer/PRO.

The most correct and fair way to allocate and control obligations with WEEE legislation is through the process Germany has implemented; an allocation of each single container to Producers/their PRO, independent of its location. As containers anywhere in the country can be assigned to a Producer, the system avoids any unfair gains by regional costs differences/distances to the recyclers.

However, reality has shown there is only a very little difference in collection/ transportation costs between rural and urban areas as the costs for pickups are not so much determined by distance but rather by time required. It may take the same time to transport a waste container from an urban collection site (e.g., due to traffic congestions) to a recycler than in rural areas with larger distances.

This allocation system established in Germany requires significant organization and financial/IT investments in online tools; used by collection sites to report containers ready for pickup, a tool to record the pickup history of Producers, and another one assigning a specific container to a Producer/informing the PRO.

Besides the administration efforts and investments in software, there have also been structural issues. When a container is collected at a collection site, the likelihood is that the (empty) container has been provided by a different PRO/waste treatment company who still maintains the legal ownership on the container itself. How could recyclers/transportation companies keep the control on their property? Who is responsible for potential damages to the container?

The German system has been close to collapse when PROs/their service providers refuse to pick up full containers where they did not deliver the empty one. This problem was resolved by an agreement between the PROs and their recyclers to move all containers into a virtual pool. The physical ownership of the container was no longer an issue.

3.4.3.3 Assignment of Collection Sites to PROs

Most countries established an allocation method based on the assignment of collection sites to a specific PRO. The PRO is supposed to pick up and treat the E-Waste customers dropped at the assigned collection sites or dealer collection points. PROs may also need to set up their own collection sites to achieve the *convenience level* required in the area they operate. The number of collection sites a PRO is responsible for depends on the collection obligation of its clients as well as the amount of E-Waste disposed at the assigned collection sites. In an ideal situation, both would match.

The amount of waste a PRO needs access to is fluctuating mainly due to changes in the number of clients they serve and their individual obligations.

A common method to solve issues caused by fluctuation of the amounts to be picked up by PROs is to assign a different number of collection sites. A PRO which collected less than obliged (under-collection) will serve more collection sites in the future and those who collected more than needed will be assigned a lower number of sites.

To fine-tune the prediction, amounts of E-Waste collected at a collection site in the past can be taken into consideration (Fig. 3.34).

Adjustment of the PRO's Obligation by changing the number of Collection Sites			
	PRO I	**PRO II**	**PRO III**
Year 1:			
Obligation	290 t	310 t	400 t
Collection Sites served by PRO	**12**	**12**	**16**
Collection	320 t	250 t	430 t
Under (- t) / Overcollection	30 t	- 60 t	30 t
Year 2			
Obligation	290 t	310 t	400 t
Under (- t) / Overcollection previous Year	30 t	- 60 t	30 t
Adjusted Collection	260 t	370 t	370 t
Collection Sites served	**10**	**15**	**13**

Fig. 3.34 Adjusting the number of Collection Sites to PROs to changes in the obligation of a PRO

However, adjusting future collections to compensate for current over-collection does have its downsides. The PRO over-collected will have already paid for the collection and treatment of the amounts. Savings due to lower collection will kick in the next period (usually one year). This is an issue especially for the PROs with ARF-based pricing, as the amounts over-collected must be financed one year in advance. In PROs with ARF pricing models, these financial issues are usually covered by increasing the safety margin and therefore the ARF charged to Producers, leading to a surplus when the initial over-collection leads to lower collection the following period.

Financial compensation by PROs who under-collect to those who over-collect is another form of compensation. The PRO who under-collected pays the PRO who over-collected a compensation sum for the E-Waste they need to fulfill the obligations of their client (Fig. 3.35).

Adjustment of the PRO's Obligation by Financial Compensation			
	PRO I	**PRO II**	**PRO III**
Year 1:			
Obligation	290 t	310 t	400 t
Collection	320 t	250 t	430 t
Under (- t) / Overcollection	30 t	- 60 t	30 t
Cost / t		$300/t	
Compensation paid by PRO II	$9,000		$9,000

Fig. 3.35 Example of financial compensation of over/under collection between PROs

Adjustments are also possible by a combination of financial compensation and the alignment of collection sites. However, financial adjustments have led to many disputes over the price to be paid for the amounts to be compensated. To avoid financial motivation to over collect, some regulators issued rules that the compensation has to be based on operational costs (not including overhead, safety margins, etc.).

▶ **Advice** Assigning collection sites/regions to a PRO may not be the most accurate, but it is the most efficient waste allocation model. Costs for managing the ongoing assignments are very small. There are countries where the PROs manage the assignment without the intervention of 3rd parties. An ideal model, but this does depend on a cooperation of PROs who are competing for clients.

To keep collection costs under control, a minimum weight/volume for a pickup should be set in the regulation to ensure efficiency (e.g., 6 mesh-boxes/pallets = 1/4 midsize truck load).

The management of collection sites/dealers prefers regional allocation as they will be served by the same company/same driver who knows their way around the specifics of the site.

With some ground rules provided by the governments (e.g., periods until under collection have to be fixed, maximum under-collection/period tolerated, index for collection and recycling costs), many PROs are managing the allocation of sites/financial compensation without the involvement of 3rd parties. Besides setting ground rules, the authorities may be asked to act as a referee where disputes between PROs cannot be resolved by themselves.

3.4.4 Costs of Clearing

The costs of Clearing functions depend on the level of outsourcing and how many of these functions remain within the remit of governments. Worldwide, the costs of Clearing vary widely and depend on the way and level of clearing.

Any Clearing functions outside the governments are directly (e.g., fees/container) or indirectly (registration fees of Producers/PROs, ARF) charged to the Producer.

It is essential that rules for waste allocation and compensation should be set in a regulatory format to apply to all stakeholders and therefore assure a level playfield.

Clearing functions are not only required in Take-Back Systems with multiple PROs. Also, with a single PRO, some forms of Clearing are necessary. They may not be visible as they are usually performed inside this PROs (Fig. 3.36).

	Multiple PROs	**One PRO** (monopoly)
Registration of Producers	Yes	Yes
Reporting of sales (weight/units)	Yes	Yes
Calculating POM	Yes	Yes
Allocation of waste (sites/containers)	Yes	*No*
Reporting of Collection / Treatment	Yes	Yes
Auditing Producers	Yes	Yes

Fig. 3.36 Clearing functions required when multiple or a single PRO offering their services

Differences in the cost of Clearing are mainly, besides organizational efficiency based on the scope of activities. For example, in Ireland, the only task of the Clearing function is to calculate, based on POM data they receive, the market share of the Producers who are clients of a PRO. The PROs negotiate among themselves without intervention of the Clearing Function, which Collection sites they serve and how they compensate over-/under collection. In cases where the PROs cannot find a mutual agreement, the authorities are moderating the process.

Ireland:	<€1/ton collected (flat fee)
Germany:	€1.5/ton collected (€7.5/container)
Illinois:	$11/ton collected (fee/ton)

In Germany the Clearing House (EAR) is not only responsible for determining the Producers' obligation. They also assign of individual containers from collection sites (when they are fully loaded) to Producers to be picked up by their PRO. In addition, they audit Producers and PROs, provide comprehensive statistics, and prepare enforcement activities.

The high cost in Illinois is caused by spreading the cost of Clearing to a low tonnage of E-Waste due to the limited scope of the legislation (TV/IT only). As the cost of a clearing function is mainly fixed costs, the more E-Waste is managed by a Clearing function, the lower the cost/ton will be.

3.4.5 Setting up Clearing Functions

Before developing the structure/organizational framework of a Clearing function, decisions have to be taken on which tasks a Clearing function should take (as described at the beginning of this chapter).

Additional decisions to be taken before such an organization is set up are:

- Government entity or set up by the PROs/Producers?
- Level of Governance.
- Representation of PROs toward authorities.
- Legal Format.
- Role of Clearing House in enforcement.

The organizational structure of Clearing functions depends on the tasks they are required to perform; if they are required to exercise public authority functions (e.g., in enforcement) and to what extent the authorities want to be involved in managing the day-to-day business.

Clearing functions which are set up as independent bodies are very often called a *Clearing House*.

There are various organizational models possible:

- The government establishes the Clearing function within a department, e.g., setting collection goals.
- The government subcontracts the Clearing function (e.g., calculating the overall collection target, the market share and the resulting obligation of a Producer) to a 3rd party experienced in dealing with confidential information (sales data) (e.g., an accounting firm).
- Producers set up/hire a body to manage clearing, example: Germany, Austria, Illinois.
- A department of an Industry Association takes on Clearing.

In a few cases the clearing infrastructure is financed by the government. In most cases, the Clearing function is financed by the Producers, either in the form of flat fees (registration fees) (e.g., Ireland) very often differentiated by the size of the Producer, or by payments per ton/container collected or both (Germany).

In Illinois, the Electronics Recycling Representation Organization (ERRO); an association of Producers, selected, after a tendering process the National Center for Electronics Recycling (NCER) as the entity to run the Clearing function for Illinois. NECR is a non-profit organization that has been supporting recycling for years.

In any case, representatives of Producers should be involved in the setup and control of the Clearing function. A group of experts from the industry could advise a Clearing function of/close to the government to ensure that decisions reflect the reality.

References

Anthesis Group/HP. (2015). The suitability of household end-of life IT for reuse in Denmark.

Arduin, R. H., Mathieux, F., Huisman, J., Blengini, G. A., Charbuillet, C., Wagner, M., et al. (2020). Novel indicators to better monitor the collection and recovery of (critical) raw materials in WEEE: Focus on screens. *Resources, Conservation and Recycling, 157*, 104772.

Atasu, A., & Subramanian, R. (2012). Extended Producer responsibility for e-waste: Individual or collective Producer responsibility? *Production and Operations Management, 21*(6), 1042–1059.

Baldé, C. P., Kuehr, R., Blumenthal, K., Fondeur Gill, S., Kern, M., Micheli, P., et al. (2015). *E-waste statistics-guidelines on classification, reporting and indicators, United Nations University, IAS-SCYCLE, Bonn (Germany)* (51 pages).

Consumer Electronics Recycling Act. 2017415 ILCS 151 Illinois, 2017. Retrieved February 17, 2023, from https://www.ilga.gov/legislation/ilcs/ilcs5.asp?ActID=3816&ChapterID=36.

Dempsey, M., Van Rossem, C., Lifset, R., Linnell, J., Gregory, J., Atasu, A., et al. (2010). *Developing practical approaches for individual Producer responsibility: Industry report to the European Commission.* INSEAD Working Paper.

Forti, V., Baldé, K., & Kuehr, R. (2018). E-waste statistics: Guidelines on classifications, reporting and indicators.

Global Electronics Council. (2015). Electronic product environmental assessment tool (EPEAT). Retrieved February 16, 2023, from https://www.epeat.net/.

Hieronymi, K., Hieronymi, J. & Faulstich, M. (2020). Nutzungsverlängerung durch Reparatur elektronischer Geräte als Strategie zur Abfallminimierung. *Müll & Abfall 11/2020* (S. 564–570). Berlin: Erich Schmidt Verlag.

Hieronymi, K. (2022). Zur Ökonomie der Wiederverwendung von Altgeräten. *Müll und Abfall 12/2022* (S. 684–689). Berlin: Erich Schmidt Verlag.

Huisman, J. V., Van der Maesen, M., Eijsbouts, R. J. J., Wang, F., Baldé, C. P., & Wielenga, C. A. (2012). *The Dutch WEEE Flows.* United Nations University, ISP-SCYCLE, Bonn, Germany, March 15.

Li, J., Liu, L., Zhao, N., Yu, K., & Zheng, L. (2013). Regional or global WEEE recycling. Where to go? *Waste Management, 33*(4), 923–934.

Magalini, F., Wang, F., Huisman, J., Kuehr, R., Baldé, K., Van Straalen, V., et al. (2016). Study on collection rates of waste electrical and electronic equipment (WEEE), possible measures to be initiated by the Commission as required by Article 7(4), 7(5), 7(6) and 7(7) of Directive 2012/19/EU on waste electrical and electronic equipment (WEEE). *European Commission (EC).*

OECD. (2016). *Extended Producer responsibility: Updated guidance for efficient waste management.* OECD Publishing.

Vaccari, M., Vinti, G., Cesaro, A., Belgiorno, V., Salhofer, S., Dias, M. I., & Jandric, A. (2019). WEEE treatment in developing countries: Environmental pollution and health consequences—an overview. *International Journal of Environmental Research and Public Health, 16*(9), 1595.

Walther, L. M. (2013). *Accruals vs. cash-basis accounting.* Retrieved September 6, 2023, from https://onlinelibrary.wiley.com/doi/abs/10.1002/9781119595984.ch2.

Wang, F., Huisman, J., Stevels, A., & Baldé, C. P. (2013). Enhancing e-waste estimates: Improving data quality by multivariate Input-Output Analysis. *Waste Management, 33*(11), 2397–2407.

Wielenga, K., & Huisman, J. (2013). (W)EEE Mass balance and market structure in Belgium.

Collection and Elements of Implementation of E-Waste

4

4.1 Operational Elements of E-Waste Legislation

4.1.1 Measuring/Controlling Success

Measuring the results of E-Waste legislation and comparing them with the targets is an essential element to ensure that the goals set are reached.

Ensuring that the data reported by all actors is also important to avoid fraud and therefore unfair cost advantage for some stakeholders.

Finally, the government needs appropriate tools/enforcing processes to ensure that not only all Producers, but also the other actors (e.g., recyclers) are following the rules established.

4.1.2 Measuring Targets

To quantify the success of E-Waste legislation, governments require regular reporting of the key targets. Collection Rates and Recycling Rates, as well as the number of Producers registered, but also violations are very common data which governments like to publish to show the success of their activities to reduce the environmental impact of E-Waste.

In the EU, the achievement of the collection target is the responsibility of the governments of the individual EU-Member State. The duty of the Producers/their PROs is to collect whatever is provided to them at collection sites. It is still debated if and to what extent Producers can change the disposal habits of their customers. To raise the awareness of the customers, most electric and electronic equipment is marketed with the "Crossed Out Wheely Bin" (Fig. 4.1).

K. Hieronymi, *Waste from Electrical and Electronic Equipment (WEEE) in Practice*,
https://doi.org/10.1007/978-3-031-40722-2_4

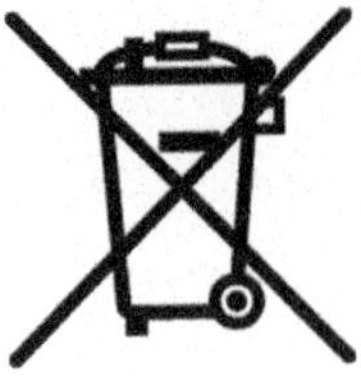

Fig. 4.1 The "Crossed Out Wheely Bin" has to be placed on all Electrical and Electronic Equipment in the EU, or, if not possible (e.g., small products) in the manual

This symbol should remind citizens to dispose of the product at the end of its use into appropriate channels. However, there is still a significant amount of Electrical and Electronic Equipment which is disposed in the *normal* household waste (up to 1.5 kg/inhabitant/year) even in countries where WEEE Regulation has been introduced several years ago. Most of the products disposed with other wastes are Small Domestic Appliances, and other small items that fit into household waste bins.

Producers have successfully argued that they cannot be made responsible for the behavior of citizens. Therefore, the ultimate measure for Producers/their PROs in most countries is that containers assigned to them/their PRO are picked up within a certain time (e.g., Germany 2 working days).

4.1.3 Controls, Audits

It is essential for the success of WEEE legislation, that there are control mechanisms in place to ensure all Producers meet their obligations. For example, did a Producer met the collection/treatment obligation?

Controlling tasks may include:

- Audits of Producers regarding amounts of POM reported.
- Audits of *private collection* activities/results of Producers.
- Quality of services provided by PROs (e.g., response time to pick-up requests).
- Auditing data submitted by PROs regarding the collected amounts.
- Auditing recycling rates provided by PROs/recyclers.
- Quality of Recycling.
- Working conditions of employees.

4.1.4 Enforcement

Enforcement of the obligations of Producers and PROs is an essential element to ensure that targets are reached and that all Producers provide their fair share.

Without proper enforcement, companies who do not care for their reputation would contribute less and therefore achieve a financial competitive advantage (=> free-riders).

Fines are the most common way to punish those who failed to fulfill their obligation. Fines have to be significant and higher than the cost savings caused by cheating the system. In severe cases, even personal liability of a Producer's management should be considered. Publishing the names of those companies who fail to comply could also be an effective tool to force Producers to comply with the E-Waste legislation.

As the ultimate punishment, import/sales restrictions should be established as they will convince the management of a Producer more than even a high financial penalty.

Clearing Functions can play a role in enforcement. This could even include to prepare cases against Producers/PROs to minimize the work of governments agencies (e.g., providing evidence). However, the final responsibility for enforcement is with authorities in countries.

The level of enforcement varies to a large extent around the globe.

4.2 Collection Categories and Product Groups

Collection Categories are established to move products with similar attributes into the appropriate recycling processes. Separating products into Collection Categories are a tool to avoid an additional process step, the sorting prior to treatment.

Product Groups are established to provide a fair allocation of the cost of collection and recycling to Producers. They help to minimize cross-subsidies between industries/Producers.

4.2.1 Collection Categories

Proper waste management requires the separation of the various wastes to ensure the appropriate recycling/disposal technology is used to minimize the environmental impact and maximize resource recovery. The first step is to separate WEEE from other waste streams. While theoretically, automated separation of WEEE from mixed household waste is possible (e.g., through video imaging separation), the preferred solution is to collect E-Waste separately.

Separate collection of WEEE avoids undetected pollution from other waste (e.g., liquids, paints, etc.) and vice versa; hazardous components from E-Waste (e.g., broken CRT-Screens, CCFL-tubes) can pollute other waste streams.

However, separating E-Waste from other waste is not all that's needed to ensure proper E-Waste treatment. E-Waste should be further separated into its own categories to ensure different electronic products receive treatment specific to their design and material content. An example for monitors/TVs is provided in Fig. 4.2.

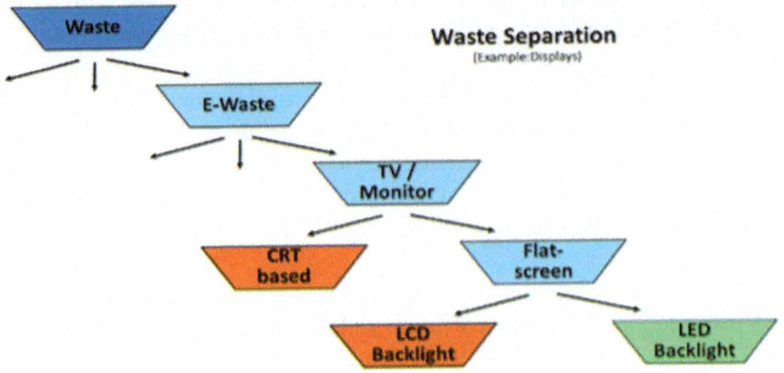

Fig. 4.2 Separation required for the proper treatment of displays

Remark: At least, E-Waste has to be separated into two treatment streams

- **Direct mechanical Processing**. A significant percentage of the E-Waste (those that **do not contain** hazardous components (e.g., CRT Tubes, CCFL lights) can be moved directly into mechanical recycling processes (e.g., shredding) followed by automated sorting of materials.
- The **removal of hazardous components**, a process step required to avoid pollution by hazardous materials like Mercury, Lead, or certain Flame Retardants is very often a manual process. Automation is tested but has not achieved the required accuracy and cost envelope. After the hazardous materials have been removed, the products are sent to mechanical treatment.

As **second hand spare parts** used in repair/remanufacturing are experiencing an increasing demand, sorting products into those who are worthwhile to be *cannibalized* is step which will provide an additional revenue potential to lower the recycling cost.

Certain products require further separation (e.g., to apply proper recycling technology). For example, TVs and Monitors with CRT, the dominant technology for TVs and monitors until the end of the previous millennium, contain leaded glass which can contaminate surface- and groundwater if not disposed correctly. CR-Tubes need to be removed from the equipment before shredding to allow special treatment, e.g., cleaning the glass from the phosphorous layer, separating glasses with different lead content and disposal of the leaded glass in special waste disposal sites designed for hazardous materials.

The first generation of flat displays used fluorescent tubes to illuminate the screen (CCFL). These tubes contain substantial amounts of Mercury and need to be removed to prevent contaminating other materials and provide a health hazard to employees in the shredding process for example.

Selecting the location where separation should be performed is a matter of cost. The more sorting is performed at a Collection Site, the more floorspace and containers are required, hence more sorting increases the collection cost. But it saves additional processes and steps at the treatment side.

The limitation of space at Collection Sites and the ability/inability of citizens to identify products for correct separation is a major reason why in many countries, products are combined into *Collection Groups/Categories.*

In case of the collection is based on picking up the E-Waste from customer sites (e.g., big housing estates), separate collection is also impossible as the trucks involved would require many separate compartments and sorting by the waste removal staff.

Transporting unsorted E-Waste to sites where they can be sorted by experts might be a solution for the collection option Curb-Side or "Milkman Tour" (Sect. 3.3.1) or for urban areas with a shortage of Collection Site space.

Categories are also required to simply the administrational processes in the context of "Put on Market" (POM). These categories, which are also called *Input-Categories,* are established to define the obligation of Producers in Take-Back Systems, based on Collective Producer Responsibility (CPR) to get closer to a *level playing field* (same impact for all Producers). Without the separation into Input-Categories, all Producers would, in a CPR-based structure, be responsible for all waste. For example, a Producer of washing machines would be accountable for the waste from TVs; a TV Producer for lamps; and a company which is importing lamps would be accountable for the waste from washing machines.

In the sense of Extended Producer Responsibility (EPR), this would be unfair, due, e.g., different return rates (e.g., the return rates from washing machines are inherently low compared to other waste, as they often end up as scrap metal) and recycling cost. The recycling cost can range from a positive income (e.g., washing machines/PCs) to high recycling costs (e.g., fridges, CRT-TVs, A/C).

Any categorization for both, POM, and collection/treatment categories, requires a delicate balance between the goals to be achieved, as well as the environmental impact (e.g., sorting, separate collection/transportation increases the carbon footprint), the administrational efforts, and finally costs.

How can Sorting into Collection Categories at Collection Sites be organized in Practice?

Citizens bring the products to collection sites and are supposed to drop products of the same Collection Category into the designated container.

Especially for smaller items (e.g., household devices, stereos) citizens may need assistance to drop the device into the appropriate container. It has been helpful to display posters with examples of products to be placed into the different containers. Closed, standard shipping containers should be preferred over open skips as they provide a level of weather protection. Longer exposure to heavy rain may de-solve some of the hazardous substances contained in electronic equipment (e.g., lead from solder) (Figs. 4.3 and 4.4).

Fig. 4.3 20 or 40 ft standard shipping containers can be used to collect E-Waste separated into product categories at municipal Collection Sites. They are preferable over open skips as they provide weather protection

Fig. 4.4 "Lattice Boxes" are perfect for segregating and transporting smaller E-Waste at retail Collection Sites. They are available in various sizes (e.g.,120 × 80 × 100 cm) and can be stacked depending on the weight

For smaller Collection Sites, e.g., at retailers, standardized *Lattice Boxes* are an alternative to help sort the E-Waste. They can be loaded and transported with standard equipment (e.g., forklifts/trucks), and the foldable versions reduce the space required when transported empty.

To assist citizens some municipalities have set up counters (Fig. 4.5) where staff is receiving the items and put them into the appropriate container. This allows municipalities to sort even more precisely, e.g., to extract rechargeable batteries and place those who are broken into specific containers to reduce fire hazards. The additional cost of these counters may be financed by separating valuable materials (e.g., cables) or receiving compensation for precise sorting from the PRO who may get lower recycling cost for avoiding further sorting for the recycling process.

Fig. 4.5 Receiving counters allow further separation at municipal collection sites

The treatment cost of monitors with CRT-Displays is usually exceeding those of flat screen monitors with LED-Backlights by far. Laptops, in many cases, receive a positive recycling value. The example illustrated in Fig. 4.6 shows the result of collecting these devices together and charging an average price. If collected in one Collection Category, Producers of products with higher cost will be subsidized by those whose products have lower cost or even a positive recycling value. In the example, e.g., the Producers of Flatscreen LED monitors will be charged the average cost of \$173/t when their products are collected and treated together in a category with screens of all technologies. If e.g., LED based Flatscreen monitors would be collected and treated separately, the cost would be only \$50/t, resulting on a cross subsidization of monitors with CRT-/CCFL displays of \$123/t. Producers of Laptops, if collected and treated together with CRT-Screens would experience an even bigger disadvantage (\$373).

Content of a Collection Category	**% of products in a Collection Category**	**Product specific Collection & Recycling costs**	**Advantage / Disadvantage**
CRT-Displays	55%	$300/t	$128/t
Flatscreen CCFL	20%	$100/t	$73/t
Flatscreen LED	15%	$50/t	$-123/t
Notebooks	10%	revenue: $200/t	$-373/t
	(weighted) Average Collection & Recycling Cost across the "Screen" Category: **$173/t**		

Fig. 4.6 Impact of combining products with different treatment costs in the same product category (e.g., "Category Screens" containing TVs, Monitors, and Notebooks)

> **Advice** The bigger the spread between the recycling costs of different product groups within the same Collection Category is, the higher is the risk of cross subsidization. This is counterproductive to the EPR-goal to reward Producers for improving their products' recyclability.

Not only different collection and treatment cost can cause an unlevel playfield. Different Collection Rates of various product categories may have an even bigger impact. It can be observed in many Take Back Systems that the Collection Rate for Notebook Computers and Smartphones is significantly lower than for screens (TV/Monitors). Many Notebooks do not end up in the E-Waste Take-Back System as they may experience a longer usage time than expected as they carry a value as a second-hand product or are cannibalized for spare parts by refurbishers.

In the example documented in Fig. 4.7, it is assumed that Notebooks have a market share of 25%, TVs and Monitors the remaining 75%. These products are collected in one combined collection group, achieving a Collection Rate of 65% (260 t). Consequently, Producers of Laptops take the responsibility for the collection and treatment of 65 t (25% of the total collection of 260 t), and those who put Monitors/TVs on the market are obliged to care for 195 t.

Producers of	Put on Market (POM)	Market Share	Collection	Obligation	Treatment Cost of Category "Screens"	Total Cost
Notebook	100 t	25 %		**65 t**	$173/t	**$11,245**
TV / Monitor	300 t	75 %		**195 t**	$173/t	**$33,735**
Total	400 t		260 t			$44,980

Fig. 4.7 Distributing the WEEE-Cost of Notebooks and Monitors to their Producers when both product types are collected and treated together

Both, the Producers of Notebooks as well as Monitors/TVs are charged with the average collection and treatment cost of $173/t (see Fig. 4.6), resulting in the bills of $11.245 for the Producers of Notebooks, $33,735 for TV-Producers.

Due to different Collection Rates, Notebooks would experience, if they are collected separately, significantly lower cost ($3,460 vs. $11,245) than in a system where both product groups would end up in one container (without separation) (Fig. 4.8) due to much lower return rates.

	Put on Market (POM)	Return Rate	Collection	Obligation	Treatment Cost of category "Screens"	Total Cost
Laptops	100 t	20%	20 t		$173 / t	**$3,460**
TV / Screen	300 t	80%	240 t		$173/ t	**$41,520**
Total	400 t		260 t			$44,980

Fig. 4.8 Calculation of cost of collecting products with different return rates separately, applying the same treatment cost/t

Cost of collection and treatment are even more favorable for Producers of Notebooks if they are not only collected but also treated separately as shown in Fig. 4.9.

	Put on Market (POM)	Market Share within Collection Group	Collection	Obligation	Treatment Cost of separated Product Categories	Total Cost
Notebook	100 t	n/a	20 t	20 t	$-200 / t	**$- 4,000**
TV / Monitor	300 t	n/a	240 t	240 t	$ 193 / t	**$46,320**
Total	400 t		260 t			$42,320

Fig. 4.9 Cost for Producers of equipment with low return rates and low treatment cost (e.g., Laptops) when these products are collected and treated separately from this with high return rates/cost

▶ **Remark** There may be different interest from different industry sectors regarding the composition of Collection Categories. Producers of products with low Collection Rates and low recycling cost will have a cost advantage if their products are collected seperately while those with higher Collection Rates/Recycling Cost might want their products to be collected together in one category.

If the number of Collection Categories is limited (e.g., due to space limitations at collection sites), preference should be given to combine products with similar recycling process/cost into one group.

If separate collection or sorting into subcategories cannot be implemented cost may be distributed in a fair way based on sampling

Advice

When setting up Collection Categories, the aim should be to minimize the differences in return patterns (recycling costs and Collection Rates) within the same Collection Category. When return patterns differ within a Collection Category, Producers of products with low recycling costs/low Collection Rates subsidized those with high recycling cost/high Collection Rates.

What might be acceptable Collection Categories (for consumer products)?

- Products containing "cooling gases" (fridges, freezers, air condition).
- Metal dominated products (e.g., LDA).
- Monitors and TVs (assuming, that the % of CRT-based equipment for monitors and TVs is similar—often CRT-based monitors have been replaced much earlier than CRT-TVs).
- PCs, Notebooks, and IT/communication devices containing a CPU (e.g., smartphones, tablets, routers).
- Plastic dominated products (e.g., SDA, power tools, personal medical devices, consumer printers, IT accessories, loudspeakers, video/ CD-players).
- Bulbs and lamps.

4.2.2 Pricing Groups

Logistics services of companies who pick up waste from, e.g., Collection Sites as well recycling are usually charged in $/ton. PROs with a pricing system based on *Amounts Collected/Treated* (ACT) can set up their bills (adding, e.g., a contribution to their fixed cost, profits) the same way to their clients, charging them by ton collected treated.

For PROs charging their clients based on Advanced Recycling Fees, the cost, charged per ton collected and treated must be converted into a fee per unit.

The example in Fig. 4.10 explains the principle on how the costs charged to the PRO are converted into an *Advanced Recycling Fee* (ARF), e.g., if 100,000 units are placed on the market—with 250 t collected and a price of $300/t, the cost per unit sold is $0,75/unit. (For simplicity, overhead, other fixed costs, and accruals for risks are not taken into consideration).

POM (units)	Collected	Cost/t	Total Cost	ARF
100,000	250 t	**$300**	$75,000	$0.75 unit POM

Fig. 4.10 Conversion of cost, charged by operators in $/t collected and treated into an Advanced Recycling Fee

As shown in Fig. 4.11, different collection and processing costs (e.g., for different products) are directly impacting the ARF.

	POM (units)	Collected	Cost/t	Total Cost	ARF
Situation 1	100,000	250 t	**$300**	$75,000	$0.75 unit POM
Situation 2	*100,000*	*250 t*	***$200***	*$50,000*	*$0.50 unit POM*

Fig. 4.11 The impact of different costs for collection/treatment on the ARF (fixed costs, to keep the example simple, accruals, etc. have been ignored)

Another factor determining the ARFs is the Collection rate. Higher Collection rates cause higher cost who trigger, distributed on the same amount of products sold (POM), higher ARFs as documented in Fig. 4.12.

	POM (units)	Collected	Cost/t	Total Cost	ARF
Situation 1	100,000	**250 t**	$300	$75,000	$0.75 unit POM
Situation 2	*100,000*	***350 t***	*$300*	*$105,000*	*$1.05 unit POM*

Fig. 4.12 The impact of an increased Collection Rate (Situation 2,350 t vs. 250 t) on the ARF (fixed costs, accruals etc. have been ignored for the simplicity of the example)

Even products of the same nature (e.g., printers) can cause different collection and recycling cost per unit. For example, the collection cost for printers who weight 2 kg (transportable Ink-Jet printer) and those whose weight is 300 kg (Office Multi-Functional Device) are very different. Even more, the recycling cost/kg can differ to a large extent as heavier products usually contain more metals than the lighter ones whose main ingredients are plastics.

To avoid unfair distribution of cost between Producers in an ARF-based pricing model, *Pricing Groups* are established, e.g., small desktop printers are put in one, bigger stand-alone office printers into another pricing category with different ARFs applied. Within a Pricing Group, e.g., of desktop printers, recycling cost might be similar, but weight and therefore collection, and transportation cost vary between products of different Producers.

▶ **Advice** Setting ARF-Pricing Groups can easily cause unlevel playfields with significant competitive advantages for some and disadvantages for other Producers.

Not only combining products with different Collection Rates, weights, and recycling costs in the same ARF-Pricing Group will cause unfair cost allocation to Producers, even elements which may be considered as of minor importance as e.g., changing the boundaries of Pricing Groups can impact the playfield of competition.

In the following example, the impact of changing the weight-scope of the Pricing Groups within the same product group (printers) will be demonstrated. Just moving the weight range of the low-end Pricing Group from 0–4 kg to 0–5.5 kg/unit in the example illustrated below will provide one Producer a cost advantage over its competitors.

The example shows the impact of small changes in setting the boundary between Advanced Recycling Fees is documented in Figs. 4.13 and 4.14. Both scenarios are based on the assumption that a Producer (Producer A) is selling three different models of equipment (e.g., printers), whose weights are 5.0 kg/unit (20,000 units sold), 11.5 kg/unit (24,000 units sold), and 13.0 kg/unit (24,000 units sold). The cost of collection and treatment for all products put on the market is estimated to be \$0.2/kg for every product POM. The resulting ARF for each Pricing Group is calculated by multiplying the (linear) average weight within a group with the cost per kg (e.g., 2.0 kg avg. weight for the Pricing Group up to 4 kg/unit multiplied with \$0.2/kg resulting in an ARF for products of the Pricing Group up to 4 kg of \$0.4/unit).

<table>
<tr><th colspan="18">Scenario I</th></tr>
<tr><td>kg/unit</td><td>3.0</td><td>3.5</td><td>4.0</td><td>5.0</td><td>5.5</td><td>6.0</td><td>7.0</td><td>8.0</td><td>8.5</td><td>9.0</td><td>10.0</td><td>11.0</td><td>11.5</td><td>12.0</td><td>13.0</td><td>14.0</td><td>15.0</td></tr>
<tr><td>Prod. A, #k units</td><td></td><td></td><td></td><td>200</td><td></td><td></td><td></td><td></td><td></td><td></td><td></td><td></td><td>24</td><td></td><td>24</td><td></td><td></td></tr>
<tr><td>ARF Sub-Category</td><td colspan="3">up to 4 kg</td><td colspan="4">4.01–7.0 kg</td><td colspan="4">7.01–10.0 kg</td><td colspan="3">10.01–12.0 kg</td><td colspan="3">12.01–15 kg</td></tr>
<tr><td>Avg. weight/category</td><td colspan="3">2.0 kg</td><td colspan="4">5.5 kg</td><td colspan="4">8.6 kg</td><td colspan="3">11.0 kg</td><td colspan="3">13.5 kg</td></tr>
<tr><td>ARF ($ 0.2 / kg POM)</td><td colspan="3">$0.4</td><td colspan="4">$1.1</td><td colspan="4">$1.7</td><td colspan="3">$2.2</td><td colspan="3">$2.7</td></tr>
<tr><td rowspan="2">Total Cost Producer A</td><td colspan="3"></td><td colspan="4">K$220</td><td colspan="4"></td><td colspan="3">K$52.8</td><td colspan="3">K$64.8</td></tr>
<tr><td colspan="17">K$338</td></tr>
</table>

Fig. 4.13 Scenario I: Calculation of an ARF for the various sub-categories based on weight of equipment. In this scenario, equipment of Producer A, weighting 5 kg/unit falls into the 2nd category with an ARF of \$1.1/unit

<table>
<tr><th colspan="18">Scenario II</th></tr>
<tr><td>kg/unit</td><td>3.0</td><td>3.5</td><td>4.0</td><td>5.0</td><td>5.5</td><td>6.0</td><td>7.0</td><td>8.0</td><td>8.5</td><td>9.0</td><td>10.0</td><td>11.0</td><td>11.5</td><td>12.0</td><td>13.0</td><td>14.0</td><td>15.0</td></tr>
<tr><td>prod. A, #k units</td><td></td><td></td><td></td><td>200</td><td></td><td></td><td></td><td></td><td></td><td></td><td></td><td></td><td>24</td><td></td><td>24</td><td></td><td></td></tr>
<tr><td>ARF Sub-Category</td><td colspan="5">up to – 5.5 kg</td><td colspan="2">5.51–7.0</td><td colspan="4">7.01–10.0 kg</td><td colspan="3">10.01–12.0 kg</td><td colspan="3">12.01–15 kg</td></tr>
<tr><td>Avg. weight /category</td><td colspan="5">2.8 kg (before: 2.0 kg)</td><td colspan="2">6.25 kg</td><td colspan="4">8.6 kg</td><td colspan="3">11.0 kg</td><td colspan="3">13.5 kg</td></tr>
<tr><td>ARF ($ 0.2 / kg POM)</td><td colspan="5">$0.56 (before $ 0.4)</td><td colspan="2">$1.25</td><td colspan="4">$1.7</td><td colspan="3">$2.2</td><td colspan="3">$2.7</td></tr>
<tr><td rowspan="2">Total Cost Producer A</td><td colspan="5">K$112</td><td colspan="2"></td><td colspan="4"></td><td colspan="3">K$52.8</td><td colspan="3">K$64.8</td></tr>
<tr><td colspan="17">K$230</td></tr>
</table>

Fig. 4.14 Scenario II: Boundaries of the low-end Pricing Group have been changed, the equipment of Producer A who weights 5 kg/unit are now in the lowest ARF-Group (0-5.5 kg) with an ARF of \$0.8/unit

In scenario I (Fig. 4.13), the low-end products of Producer A fall into the 2nd Pricing Group (4.1–7 kg) with an ARF of $1.10/unit based on the average weight within this group of 5.5 kg/unit (times $0.2/kg). Adding the ARFs Producer A must pay for all his mid-range and high-end products, his total bill for the collection and treatment of his all of his products in all 3 pricing groups is K$337.6.

Changing the parameters of pricing group slightly can have a significant impact on the bill of an individual Producer, without changing any of the Collection Rate, Recycling Cost etc. In Scenario II (Fig. 4.14), the boundary of the low-end Pricing Group (formerly up to 4kg) has been moved to include products with a weight of up to 5.5 kg/units., As a result, the low-end products of Producer A fall into now into the new low-end group where the ARF is with $0.56/unit, lower than in the previous scenario ($1.10). At the same time, Producers whose products have been in the Pricing Group with an ARF of $0.4 in Scenario I will face an increase from $0.4 to $0.56/unit in Scenario II.

► **Remark** By just changing a small element in the definition of a Pricing Group, the total cost for Producer A will decrease by approx. 1/3.

On the flip side, Producers who's equipment weights less than 4 kg/unit will see a significant cost increase.

As indicated in Fig. 4.6–4.9 the Collection Rate is another very important cost factor to consider. Different products may experience different collection rates, e.g., products with higher quality are used longer and/or are exported as functional 2nd hand products. Producers of high-quality products mixed with low-quality products in the same category are subsidizing the low-quality Producers. The Producer of products with a shorter usage time pays the same ARF, but the amount of waste to be collected and processed from low-quality products is higher.

Variations in recycling cost should be considered when defining Pricing Categories for which the same ARF is applied.

The following table contains a real-life example from the EU who mixed high-cost products (TV Screens) with products with a positive recycling value ($−200/t) into the same category, resulting in Producers of Laptops cross subsidizing TVs.

Figure 4.15 shows the (simplified) calculation when Laptops and TVs are combined in one pricing category, resulting in an ARF of $1.09/unit, both for TVs and Laptops.

	Sales (units)	Amount collected	Collection & Treatment Cost	Other cost (e.g., admin, risk prevention)	Total Cost	Cost / Product (ARF)
Laptop	10,000	20 t	$-200/t	$35/t		
TV	50,000	240 t	$250/t	$35/t		
Total	60,000				**$65.100**	**$1.09**

Fig. 4.15 Laptops and TVs included in one Pricing Group, ARF based on the average cost of both types of products

Figure 4.16 shows the impact, if separate ARF categories are established for Laptops and TVs. Without any changes of the operational parameters (Collection Rate, treatment cost), the ARF for TVs will increase. Producers of Laptops should be paid $0.33/unit to reflect the positive recycling value of these products.

	Sales (units)	Collection	Treatment Cost	Other cost	Total Cost	Cost / Product (ARF)
Notebook	10,000	20 t	$-200/t	$35/t	**$-3,300**	**$-0.33**
TV	50,000	240 t	$250/t	$35/t	**$68,400**	**$1.37**
Total						

Fig. 4.16 Different ARF Pricing Groups established for Laptops and TVs, ARF based on specific cost of each group

> **Remark** Recycling revenues can be different by country/region and change over time due to labour rates and dependencies on the commodity market. Positive recycling values may not be achieved everywhere, especially in countries with low collection volumes.

The Belgian PRO (RECUPEL) established only one ARF Pricing Group for displays (Monitor, TVs), independent of their weight/size. As large monitors (e.g., 70") weigh approx. >6 times more than a smaller one (e.g., 12"), small monitors are subsidizing large screens. Considering that 6 small monitors' weight is the same as the weight of one big screen, the small monitors pay $10.20 compared to the ARF for a big screen of $1.70.

To minimize the cross-subsidies between ARF Pricing Groups for monitors, the Swiss PRO (SWICO) launched five ARF Pricing Groups for the monitor category, based on a monitor's size (assuming that size is an equivalent to weight/processing cost).

	Swiss ARF *	Belgian ARF**
<15"	$1.75	$1.55
>15"	$5.20	$1.55
>33"	$10.50	$1.55
>45"	$16.71	$1.55
>70"	$24.50	$1.55

Fig. 4.17 Comparing the prices for monitors of the Swiss and Belgian PRO. * net prices w/o sales tax (2021) ** tariff from Jan 2022 onwards

Setting up ARF Pricing Groups will reduce but not necessarily avoid cross subsidization between products. In Switzerland where different ARFs are set for screens according to the size of monitors, cross subsidization, e.g., will occur within an ARF Pricing Group: 32" monitors are charged the same ARF as a 42" screens, even though the collection and recycling costs of the bigger monitor is higher (due to higher weight).

Setting up fair ARF Pricing Categories is even more complicated when the specific Collection Rate is included to the criteria of Pricing Groups. Very often, the quality of products determines the usage. High-quality products do show a lower Collection Rate as they are used longer; an attribute that should be rewarded according to the EPR principle. Determining the Collection Rate for a specific product (group) or brand requires sampling and significant extra effort. Including the Collection Rate into the composition of pricing groups is not practical, given the massive variety of combinations involved.

Some PROs are charging an ARF not based on the number of products but on their weight of the new products put onto the market. This system avoids the setup of weight-based pricing groups as the ARF for every product is individually calculated based on its weight. In such a system, the PRO is setting an ARF per kg for a specific product category, reflecting the collection and recycling cost specific to this product group. For the examples illustrated in Figs. 4.13 and 4.14, the ARF for a product would be set by its actual weight and not an average of the Pricing Group. This way will level the playfield and avoids unfair cost distribution caused by the setup of Pricing Groups.

▶ **Remark** Setting fair ARF Pricing Categories/Groups requires a delicate balance. The establishment of many Pricing Categories, taking weight, Collection Rate, and treatment cost into consideration will reduce cross-subsidies between Producers. But they increase the administrative efforts of Producers and even more those of the PROs.

Selecting a pricing method based *Amounts Collected/Treated* (ACT), where Producers pay for the waste allocated to them and collected/treated by the PRO avoids at least the weight-based cross subsidization.

There are even more challenges when categories are based on data which require efforts to extract them out of company systems. For example, in Austria, in the first WEEE Implementation, product categories have been based on the physical dimensions (length, width, height) of a product. This data is usually not recorded in databases used by sales organizations of producers as it is not relevant for the sales/distribution process. Consequently, individual products had to be assigned to categories using a manual process (e.g., checking product datasheets on individual products), an activity which must be performed not only once but monthly for the new products leading to increased compliance cost, especially for Producers who offered a wide range of products.

In the meantime, Austria did follow the categorization established by the EU, using 6 categories where market share by weight is used to allocate the E-Waste obligation of the individual Producer.

4.3 Producer Responsibility Organizations (PRO)

PROs are important an important element of Take-Back Systems. They are managing the operation of collection, transport, and treatment on behalf of their clients.

4.3.1 Role of Producer Responsibility Organizations (PROs)

The academic principle of Extended Producer Responsibility is based on moving the responsibility for the End-of-Life phase of a product to the Producer. Almost all Producers do not have the skills and the resources to manage the collection, the recycling/disposal processes, and the optimal operational setup to ensure proper collection and treatment of the waste they are responsible for. Consequently, Producers outsource these services to specialized partners.

Organizations that perform the operational obligations of the Producers caused by local E-Waste regulations are called **Producer Responsibility Organizations (PROs)**. PROs can be either dedicated organizations or belong to another company (e.g., subsidiaries of waste management companies, recyclers) who are performing the operational elements of the Producer's duties, with regards to EPR-based legislation.

Depending on the national E-Waste laws, the regulations that the setup of PROs must follow can vary widely:

- **Almost no rules**. Example: in Germany, PROs just need to carry a generic waste management license.
- **Rules specifying the legal format**. Example: in Italy, being associations of Producers is required.
- **Limited type of shareholders,** e.g., Producers only
- **Minimum capital of the company**
- **Composition of the Board**. Example: France requires a government representative on the Board.
- **PROs need to be a national enterprise** (registered in the country where they operate) (e.g., Italy).

While some countries also require these organizations to be Not-for-Profit (e.g., South Africa), others encourage commercial companies (often from the waste logistics sector or recyclers) to act as a PRO (e.g., Germany, UK).

In addition to regulatory differences in the setup/organization, essential differences can be observed which responsibilities could be transferred from the Producers to the PRO. In many countries, including the UK and Germany but also South Africa, the overall E-Waste management responsibility remains with the Producers. When the PRO fails to deliver either the service quality (e.g., prompt pick-up) or proper waste treatment, the Producers are still responsible. This is in line with the Extended Producer Responsibility Principle, ensuring the Producers fulfill their E-Waste obligations.

In a few countries (e.g., Austria), the legal obligations for collection and treatment are transferred from a Producer to its PRO. In these systems, Producers must be a client of a PRO, report the appropriate input figures, and pay their bills. Any other non-compliance elements (e.g., failure to collect) are transferred and enforced with the PRO. While this can be seen as a favorable structure, there are also risks for the Producer. In a system where the obligation stays with the Producer, a PRO can stop collecting E-Waste in case a Producer is not paying his bills. Some countries, e.g., Austria make the PRO responsible for the compliance of their clients. E.g., if a client failed to pay their bill, the PRO carries the responsibily of this Producer until the end of a reporting period

> **Remark** In cases where the obligation of Producers has been moved to the PRO, they need to collect and recycle the waste of all clients until the end of a reporting period, regardless of whether all Producers paid their bill or not. As a result, all other clients/members of a PRO will be paying for the obligations of the Producer that did not pay its bills.

The involvement of Producers in managing PROs is very different and depends on the local setup and the legal requirements. Some countries, like Italy, request that PROs are set up by Producers, who retain a financial stake and participate in the organization's management.

The following sub-chapters describe various elements for setting up PROs. This list is not complete but describes significant factors to consider when setting up a PRO.

4.3.2 Operational Structure of PROs

4.3.2.1 Government Organization

There are a few countries, among them China and Taiwan, where the government organizes the operation of WEEE legislation. Government departments distribute the funds collected from Producers to pay for the recycling and/or subsidize treatment. Collection is neither regulated nor supported financially (Salhofer et al., 2016). Thereby the whole collection of WEEE is based upon the actions of the (more or less) informal sector (Streicher-Porte & Geering, 2010). There is no further responsibility for Producers in these systems other than to provide funds (e.g., on a fee/POM basis).

4.3.2.2 Single PRO

Countries/sectors may choose to set up a single PRO. Producers have no choice in these cases and are forced to use this PRO. Take-Back Systems with a single PRO are sometimes called "Monopolistic Take-Back Systems" (e.g., Belgium, South Korea, Switzerland)

In Switzerland, e.g., there have been 3 PROs, one for domestic appliances, another one for IT and Consumer Electronics (CE), and one for lamps, providing take back services for their industries' specific products.

SWICO Recycling is part of the Industry Association of IT and Consumer Electronics. SENS is a foundation initiated by the Producer association for domestic appliances. At the same time, SLRs is the PRO founded by the association of Swiss Producers of lamps. SENS and SLRs cooperated since quite some time (e.g., SENS-Collection sites accepted lamps) before they recently merged into one organization.

Even though there are multiple PROs operating, Producers of a specific industry sector have no choice but to use the only PRO operating for their products. As Producer do not have a choice in which PRO they want to use, even Take-Back Systems were more than on PROs are in operation can be called *Monopolistic*.

4.3.2.3 Multiple PROs

In a TBS with multiple PROs in operation, Producers can choose the best offer from the PRO, which fits their specific requirements. PROs do not only differentiate their services through the price they charge. As in any competitive system, they are eager to tailor their services to the requirements of their clients. For example, several PROs offer both pricing methods Advanced Recycling Fees (ARF) for clients who are not interested in digging into E-Waste management details as well as prices based on the Amounts Collected/Treated (ACT) pricing model for those who are able to use Take-Back to their advantage. PROs often expand their offer to services outside of legal compliance in conjunction with 2nd hand products (e.g., data cleaning) or even provide access to recycled materials for Producers.

> **Remark** Especially for the government it is easier to communicate with a single PRO than with many. However, a market-based framework, allowing more than one PRO to provide services will provide the Producers convenient solutions for their specific needs.
>
> In many countries with multiple PROs, the **Clearing function can serve as a single interface between the authorities and the Producers.**

In a TBS with multiple PROs, the Clearing function plays a central role in bridging a potential communication gap between Producers, the PROs, and the authorities.

Germany is an example where the interface between the PROs, the Producers, and the government regarding the operational aspects of the E-Waste legislation is performed by the Clearing House (Elektro-Altgeräte Register (EAR)). EAR helps the Environmental Agency in Germany to enforce WEEE legislation by preparing legal cases against violators, that the authorities can execute without further preparation. They are also the focal point for discussions between the industry and the government, e.g., regarding measures to increase the collection (promotion campaigns, etc.). With this model, Germany's enforcement agencies have a single interface with the industry even though approx. 30 PROs are in operation.

Almost all clearing functions are required in both, Monopolistic Systems and those based on competition between PROs. The difference is that a separate Clearing function has to be established in *Multi-PRO* systems while in a monopolistic structure Clearing it can be integrated in the activities of the single PRO (for details see Sect. 3.4).

Allocation of waste between PROs is certainly a function only required in a competitive TBS. The costs of this activity depend on the way waste is allocated to individual PROs. For example, in Germany, where individual containers are allocated to Producers/their PRO, the costs of the entire Clearing Function are approx. $1.5/ton (collected). In Ireland, a TBS where Clearing is based on the assignment of Collection Sites/regions to a PRO, the Clearing costs are charged as a flat rate of $2,000/year (for large companies). Clearing costs in both countries for a Notebook put onto the market result in <$0,01 (for a large international Producer).

Economy of scale is an essential factor for the cost of waste management. The higher the amount collected/treated, the lower the costs charged from transportation companies and recyclers (on a per t basis). PROs operating across borders can easily achieve the same (or even better) economy of scale effects than monopolistic systems who are operating within a country. PROs operating in multiple jurisdictions can accumulate larger waste amounts across several countries, and use the price differences in the countries, e.g., for sorting/recycling. They also can share their fixed costs across countries. PROs operating across borders cannot only be found in Europe and the Americas, but also in Asia where a German PRO is operating in Hongkong and Singapore.

Also, competition between PROs leads an economic motivation to reduce costs (e.g., fixed costs), a motivation which is, to an extend, missing in some monopolistic structures. Reduced waste management costs are not only benefitting Producers, but also the customer. Independent of fees or other financial models, the customer will pay the bill for high Take-Back Cost. Producers report that PROs in competitive TBS are more responsive to the special needs of their clients (e.g., reports and additional services) than in countries where there is no choice but the national monopoly.

Figure 4.18 shows the price developments in Austria in 2004; the year WEEE was implemented there. With approximately eight million inhabitants, Austria is one of the smaller countries in the EU.

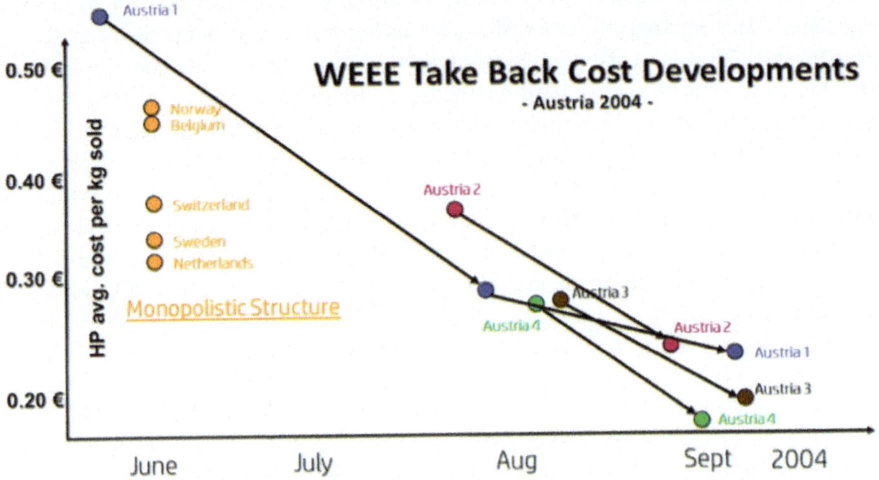

Fig. 4.18 Price developments for WEEE in Austria before implementing WEEE legislation (Sept. 2004) and a comparison with monopolistic systems in other countries for the same period

In June, a PRO (Austria 1) announced their prices of €0.6 (printer category) for every kg sold (POM). At the end of July, a second PRO (Austria 2) gave a price of €0.38/kg sold, followed by Austria 3 with €0.29.

When WEEE started in autumn 2004, prices dropped down to a level of €0.18/kg, less than 1/3 of what had been asked a few months before and much lower than monopolistic systems charged in countries of a comparable size (as illustrated in Fig. 4.18).

There are Monopolistic TBS with very low prices. RECUPEL, the single PRO for E-Waste in Belgium charges very low prices. Besides a comparable low Collection Rate, RECUPEL has been using the reserves they built in the early years of WEEE to subsidize their fees. But reserves will come to an end. Their pricelist (Appliance List Jan 1st, 2022) which they recently published, shows prices increased (over 2020) by approximately 100% (e.g., for monitors). But even with these price increases, they offer very competitive prices, in some categories the lowest across the globe. One of the reason that lead to the very competitive prices in Belgium have been the strong involvement of representatives of Producers in the management of RECUPEL.

Many PROs focus on one waste stream, e.g., packaging waste, E-Waste, waste batteries. Several PROs, especially those operating in multiple countries, offer compliance services for more than one waste stream.

4.4 Set-Up of Producer Responsibility Organizations (PRO)

Setting up a PRO is a complex undertaking. Many elements should be considered, starting with the legal format, the pricing model, and staffing the functions needed within the PRO. The fundamental goal of a PRO is to ensure compliance for its clients at the lowest costs, while providing an effective interface for Producers and the enforcement bodies of a country. It should be considered that a PRO is not perfect right from the beginning; it needs improvements/changes, especially in the first years.

4.4.1 Legal Format

The legal format drives many developments on setting up a PRO and should be carefully selected. It must be acknowledged that some enforcement agencies are prescribing not only the legal format (e.g., in Italy: association) but also the member/ownership and the composition of the board.

4.4.1.1 Company

The easiest way is to set up a PRO in the form of a **company**. It can be initiated by a single stakeholder (or a group of stakeholders) within a short timeframe. Using an existing company as the basis is another path to take. In the case that a single PRO is serving the market, emphasis should be placed on representing all Producers, e.g., advisory council, role, and power of shareholders. If a TBS is based on competition, the representation of clients will not be as crucial as in a monopolistic system. In a competitive system, PROs will adjust their service offerings toward clients' needs caused by market forces.

In the early days of WEEE in Europe, when no PRO operated across borders, some Producers from different product sectors funded the "European Recycling Platform" (ERP), the first Pan-European PRO for WEEE. After five years, when ERP has been stable, the Producers who owned ERP sold their shares to a waste management company who is successfully continuing to grow this PRO.

4.4.1.2 Other Forms of Organizational Setup

Some countries have regulated the legal format of PROs, to demonstrate the responsibility of Producers according to EPR. For example, Italy requests that PROs are set up as **associations of manufacturers**. To keep cost under control by sharing resources across countries and to limit the financial risks for the associates, some PROs in Italy are outsourcing the operations to (international) companies who are experienced in running the operations of PROs in other countries.

Many PROs have been initiated by Producer's **industry associations**. Depending on country legislation and the legal format of industry associations, they may need to set up a separate entity to run an operational business. In case the PRO is set up as an association, special attention needs to be put on the liability of the members of the association. It could well be, that the associates carry all financial risks of the PRO, a situation which could be avoided by using a different legal format (e.g., LTD).

4.4.1.3 Who Should be the Legal Owner of a PRO?

There are no restrictions on a PRO if a commercial setup has been chosen. Companies active in other areas of waste management (e.g., household waste collection, batteries, packaging, recycling) expanded their scope into E-Waste. Independent from who owns these enterprises, market mechanisms will ensure they provide services to the benefit of their clients.

Care must be taken in monopolistic systems, where market mechanisms are not helping to control a PRO. Producers who are formal stakeholders/owners of a PRO can influence the direction of the PRO but also carry financial risks. However, efforts of Producers' representatives required in the management of a PRO can be huge, especially in the starting phase and are very often underestimated.

In a competitive environment, involving other stakeholders of the Take-Back process, (e.g., waste logistic companies, recyclers) as the owners/founders/members of a PRO are not an issue. They will be forced by market pressure to serve the interest of their clients. In monopolistic PROs, owners/membership of other stakeholders in the waste management supply chain is not recommended. They may be able to tune the PRO toward their interest. The ultimate weapon of a Producer when he is not satisfied with its service provider, to choose a different supplier and leave the PRO, is not possible in a monopolistic PRO.

Some countries request a government representative on the board of the PRO (e.g., France). If the government representative is an observer only, it provides a chance for government's entities to better understand operational issues and the possibility to discuss potential solutions informally. But very often, representatives from the government do not have the experience in managing enterprises with a multi-million budget.

4.4.2 Profit/Not-For-Profit

Some countries require PROs to be Not-for-Profit organizations. This condition doesn't always lead to low prices for Producers and in consequence to national consumers. Profits can be made at other steps along the (waste) supply chain, not necessarily at the PRO level. The best way to avoid excessive profits is to allow for competition among PROs, limiting the profits through market mechanisms.

4.5 Inside a PRO

4.5.1 Management Structure of a PRO

PROs require similar departments to any commercial company or association. The closer a PRO is about to start operations, the more critical the procurement function is, as many functions can be outsourced. In addition to these traditional departments and customer relations, a PRO requires strong legal support. Due to the extreme dependency on the interpretation of the respective laws, a PRO needs access to knowledge of the authorities' particular interpretation and enforcement practices. Consequently, the initial skills required at the beginning are from financial, procurement, IT, government relations, and legal. Especially in smaller countries, some of the functions can be combined (e.g., legal and government affairs) to keep the organization lean (Fig. 4.19).

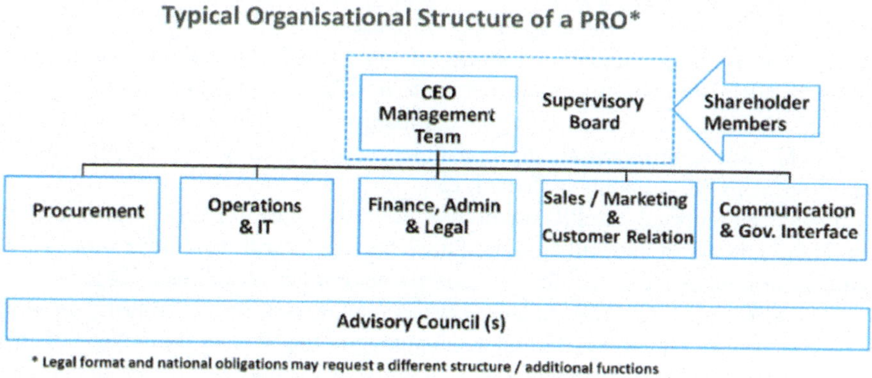

Fig. 4.19 Typical setup of a PRO

► **Advice** It has been beneficial to set up advisory councils on specific topics within a PRO, consisting out of experts from Producers, recyclers, collection (municipalities), and academics. These advisory councils can advise the PRO to take the right decisions and serve as expert groups for certain topics (e.g., logistics, pricing, recycling strategy).

► **Remark** On average, it takes 12–18 months to start a PRO from scratch. Much more time than asking an already established PRO (e.g., from another country) to establish a subsidiary.

4.5.2 PRO Staff

Hiring the CEO is the most urgent and important task when starting a PRO. It is wise to start the selection process for this job before finalizing the legal setup of the organization, so the CEO can start immediately after registration or can support the setup.

The profile of a CEO in the starting phase of a PRO is quite different as when the company is in a more mature stage. An entrepreneurial profile and solid knowledge on the financials of a company are required to ensure a solid start of a PRO, while later process optimization and focus on customers are more important.

But as with every company, financial management is an important role to fill, especially in the starting phase of a PRO (to ensure competitive prices and solid financials).

In case, a CEO comes with solid financial knowledge, CEO and CFO functions could be combined.

PROs run a complex operation that is mainly based on outsourcing operational functions to suppliers. Besides proper procurement and supplier management skills, IT is an important pillar for the efficiency of PROs. IT can be the basis to ensure adequate communication between the Clearing Function, Municipal and other Collection Sites, transportation companies, and the recyclers. To a large extent, IT also contributes to the convenience of reporting and documentation, reporting of POM, amounts collected, etc.

It has been reported that PROs who recruited their staff from companies (including clients) based on the skills required have been better off than those, whose staff came from industry associations or even the government (exceptions apply).

A good practice has been to send retired managers from the industry to work for a few years in a PRO. The major advantage is that former employees of Producers speak "Producer language" and know what is important for Producers. So, communication is much easier with former Producer employees than with people with a background in associations or politics.

Producers should compare Take-Back costs in different countries and confront PROs with their findings. Even when the PROs do not react immediately, pressure (even sometimes through the media) ensures even PROs who don't have any market pressure, will be more cost sensitive to their efficiency.

4.5.3 Producer Governance

In a TBS based on competition between PROs, market forces will ensure that PROs act in the interest of their clients. Producers have a choice to leave a PRO who is not fulfilling their expectation... .

In a monopolistics sitaution, where Producers do not have a choice between PROs, it is essential to set up appropriate structures to ensure alignment between the clients and the PRO. In several instances, monopolistic PROs allow Producer's representatives to join the Board, which is supposed to control the management and direction of the PRO.

PROs may establish customer advisory councils to ensure proper communication with their clients. When setting up an advisory council, it should be ensured all industry segments served by the PRO are represented. The council's composition may reflect the size of the industry segments the PRO serves.

Reflecting the different needs of different Producers, especially if they are active in different market segments, is a complex undertaking for industry representatives and can become time consuming. It should be ensured the valid interests of individual Producers, or Producers of a specific product category, are recognized and seriously considered in the PROs decisions.

Producer Representatives should be involved in decisions regarding:

- Procurement Guidelines.
- Quality Criteria.
- Definition of Pricing categories.
- Pricing guidelines.
- Establishing and dissolving accruals/reserves.
- Additional service offering.
- Staffing leading functions.

▶ **Remark** PROs in a competitive Take-Back System require significantly lower involvement of Producers than those in monopolistic schemes. The efforts to control the PROs in monopolistic schemes and to ensure that they understand the needs of Producers can be significant undertakings. Not all Producers do support the setup of competitive PROs. This can be observed very often with Producers who have limited experience in waste management. Governments are even more challenging to convince them toward a competitive structure. They fear that dealing with several players is more challenging than interfacing with a single PRO. The Clearing function can be a single point of contact for the (e.g., in Germany) government, providing all necessary information and enable a two way conversation between the Producers and the Government

4.5.4 Composition of a PRO-Board

Board members of companies are usually chosen by those who have a financial interest in the PRO, e.g., shareholders. It should be ensured that the composition of the Board of a PRO reflects the structure of the members/clients. An example is ANZRP, one of the Australian PROs who has been founded by large Producers.

Three of the PRO's board seats are reserved for large clients (A-members), three for smaller clients, and three for experts on topics important for the PRO (e.g., legal, marketing, lobbying). While decisions, based on the constitution of the PRO are made based on majority, the common practice established in the past years has been unanimity. Convincing each other is an appropriate way to ensure, that the interest of all clients is met, but it may take quite some efforts/time.

There are PROs who limit Board membership (and therefore the voting) to those Producers who founded the PRO. This is justified because these members finance the setup of a PRO. Any change of the market structure (e.g., changes to the size of Producers) cannot be reflected in a board composition of *founders only*. A compromise may be that a certain number of board seats are reserved for founders, but newcomers also have a chance to get a seat on the Board.

Reserving seats for certain industries (e.g., according to the Producers' contribution to the revenues of the PRO) is one way, PROs are ensuring proper representation of its members. Those Producers who do not have a formal stake in a PRO, can either be included by reserving a board seat or through a less powerful advisory council.

The format chosen for the client/member representation depends on how much effort Producers would like to spend on participating in the management of the PRO. The Producer activities may be high at the beginning of WEEE implementation and lower once a PRO is in a stable situation.

4.5.5 Decision Making

Besides the composition of the board itself, voting rights within the Board (or the general assembly) are an essential tool to ensure the interests of individual Producers/Producer groups are reflected. The usual rule in boards is, that decisions are made by the majority of the members of the Board. Such a rule, "one company, one vote", carries the risk that decisions can be taken in the interest of many small companies who do not reflect those of a few bigger corporations (or vis a versa).

There have been consortia where the majority of the board members votes against the interest of two bigger companies who provided 2/3 of the budget.

Voting rights based on the budget contribution will benefit the big players. There are a few solutions that might have eased the situation/would have increased the pressure on Board Members to find compromises (Fig. 4.20).

- Decisions made based on unanimity. As mentioned, this can be quite time consuming and may at the end block important decisions as one board member can hold a decision. The way ANZRP operates (if consensus cannot be reached, a majority-based decision is possible) seems to be a good compromise. However, this is only possible if the members of the board have similar views and direction. Especially when the PRO is serving Producers from different industries, fundamental differences in the interest of the members may be observed.

- Establishing a majority level of 60% (or higher) of the votes required is another possibility to protect minorities.
- Establishing a *double majority* is another (even better) solution to balance the interest of smaller and larger members. In a *double majority*, a decision is made, when most of the members and the majority of the budget contribution/market share agree.

The following example shows how a double majority could work in practice (based on Fig. 4.20):

Company	Budget Contribution
A	42 %
B	24 %
C	13%
D	11%
E	10%
Total	100%

Fig. 4.20 List of companies and their relative share of contributions to the PRO budget as a basis of decisions

If companies A & B support a proposal, and companies C, D, and E are opposing it, the proposal is denied. In this case, the majority of the budget contribution has been reached (66%), but not the majority of the members agree (2 in support, 3 disagree). If the proposal is edited and then supported by A, B, and C, the majority of the budget contribution (76%) has been reached, and the number of companies supporting the decision exceeds (3) those who oppose it (2), the proposal is agreed.

4.5.6 Producer Representation in the Board of a PRO

Experience shows that representing a Producer in a PRO can be an almost full-time job, at least at the beginning. Meetings have to be prepared and the impacts of decisions on Producers have to be evaluated, supporters to be convinced, and compromises to be found. As many decisions have to be taken surrounded by unknown territory, finding and getting support for a concession is time consuming.

4.6 Managing a PRO within a Producer

4.6.1 Local or Central Management of WEEE

Basically, Producers have three options: Either to manage their E-Waste obligation with local staff or establish a team of specialists who manage this topic on a global basis.

Managing E-Waste locally might not require full-time experts, the job is often done by employees dealing with that topic on a part-time basis, very often from procurement or administration departments with no or only limited experience in managing E-Waste. Managing WEEE-Cost in a department which carries a regional or global responsibility allows to employ waste experts with specific procurement knowledge for waste services. Other than in companies where Take-Back costs are managed locally, central control of Take-Back costs allows transparency of the costs differences and provides valuable input into increasing the efficiency of PROs across countries.

Different accounting principles of companies may also have an impact on where Take-Back is managed. In those companies where Take-Back costs are considered *Local Sales Costs*, the responsibility lies in the hands of the local management team.

Within a growing number of companies, Take-Back costs are considered a manufacturing cost. This is aligned with the principle of EPR. It ensures that these costs are visible in the Producer's production cost calculations. If accumulated across multiple countries, it can be significant enough to trigger changes in product design while when it is reported as local sales cost.

4.6.2 Producer Representative in a PRO

PROs very often prefer representatives from high-level management of Producers to represent their members. The downside is, for example, a company represented by its CEO who is not familiar with the specifics of waste management and may not have the time to prepare, especially if ad-hoc decisions have to be made during the meeting. The same applies to other high-level management positions, e.g., a F & A manager. The costs of sending a high-level manager is a factor to consider.

On the other hand, environmental/waste experts will understand in detail what decisions are about and what consequences they will have on their company. However, not all of them will have the background to take appropriate decisions, e.g., on financial matters.

▶ **Advice** Experience has shown that Environmental Waste Experts with some experience in business management, are good representatives on advisory councils or Boards of PROs. It should be ensured they have access to different functions (e.g., the finance manager) of their company to get advice and direction on areas, where their experience is low.

In this context, it is vital that the management and important members of its staff have appropriate skills in the English language. This will not only allow them to communicate and exchange experience with other PROs but also read the important papers who are all published in this language.

▶ **Advice** PROs are important organizations who will ensure the success of E-Waste legislation. A framework where multiple PROs offer their services to Producers, coordinated by a Clearing function and proper enforcement tools, will ensure that the goals of a legislation based on Extended Producer Responsibility are maximized at the most efficient cost. PROs operating across countries will bring huge benefits due to the experience gained in the past. There may be situations where a monopolistic PRO can be the better choice (e.g., on small remote islands). Allowing Producers to subtract their own collection and treatment activities (e.g., thru sales incentives) may provide a form of competition even there.

References

Salhofer, S., Steuer, B., Ramusch, R., & Beigl, P. (2016). WEEE management in Europe and China–a comparison. *Waste Management, 57*, 27–35.

Streicher-Porte, M., & Geering, A. C. (2010). Opportunities and threats of current e-waste collection system in China: A case study from Taizhou with a focus on refrigerators, washing machines, and televisions. *Environmental Engineering Science, 27*(1), 29–36.

5 Exports of E-Waste

Besides labor, cost of E-Waste treatment depends to a large extent on the environmental standards used in the recycling operations and handling the leftovers from this process. In proper recycling, hazardous materials are separated, emissions are avoided, and the hazardous materials are either destroyed (e.g., thru incineration) or disposed in special waste sites.

The more hazardous components or non-recyclable materials a product contains, the higher is the difference in cost between proper and improper treatment. TVs/monitors operating with CRT technology are, besides Cooling/AC devices, currently the products with the highest treatment cost. Even if almost no CRT-based TV-Set/monitor is sold today, they are still returned in significant quantities. In proper recycling, major cost is caused by the disassembly of the CRT tube (very often in a manual process) from the rest. The tube is then dismantled to enable the removal of copper-based coil. The fluorescent coating is washed from the front glass and the different types of leaded glass who make up the tube are separated. In most cases, the leaded glass and the coating are stored in disposal sites who are designed for hazardous materials (e.g., avoiding contact with groundwater, distribution of contaminated dust).

In improper treatment, to its extreme, only the coil is removed, the rest is disposed in illegal waste dumps, rivers, or the country site, leading to a positive income for those who act against the basic principles of waste management.

Significant differences in the cost of treatment let, especially in the early days of compulsory E-Waste management in Europe and North America, to significant amounts of waste exported to developing countries, leading to severe damages of the local environment and danger to the health of people involved in these activities or living near to the illgeal recycling operations

Especially shipping E-Waste to China has been very attractive for the illegal actors as the cost of transport has been very low. Even with longer distances,

K. Hieronymi, *Waste from Electrical and Electronic Equipment (WEEE) in Practice*,
https://doi.org/10.1007/978-3-031-40722-2_5

freight rates to China have been lower than e.g., to Nigeria. As empty containers, where goods from China had been transported to the western countries had to be shipped back to China anyway, transportation companies offered very low rates for freight originated in Europe/North America with the destination China.

Environmental NGOs, namely, the *Basel Action Network* identified the issue and traced E-Waste from the US/Europe, even municipal collection sites, ending up in Asia.

Their activities supported an import ban of E-Waste by the Chinese government, leading to complete import ban for all wastes recently.

Along the proverb "E-Waste finds its cheapest resting place", African and Asian Countries countries have been also the destination for E-Waste from industrialized countries (Pucket et al., 2005), mainly to western Africa (e.g., Nigeria, Ghana, Malaysia).

With the increased enforcement of export regulations in industrial countries, the implementation of new treatment technology, and increased transportation cost, exports of E-Waste to these countries have dropped significantly in the past years.

However, illegal E-waste treatment is still an issue there. According to a study of the United Nations Environmental Program (UNEP), up to 85% of the E-Waste in West African Countries is domestically generated (Schluep et al., 2012) out of the consumption of new and used equipment.

Another result of this study has been that approx. 70% of the imports were used electrical and electronic equipment. Traders, very often originated in African countries, import used equipment from all around the globe (Europe, USA, China) (Figs. 5.1 and 5.2).

Fig. 5.1 2nd hand electronic products collected and ready for shipment to African countries. Pictures taken in the Mongkok District of HongKong (2009)

Fig. 5.2 Treatment of e-waste, very often illegally imported into developing countries creates hazards to the environment and the health of workers and bystanders

Most of these products are not tested before shipment. While 70% of the shipments could be sold for immediate re-use with a reasonable lifespan, 15% require repair (e.g., to fix minor damages), but 15% had to be scrapped and are therefore E-Waste (Schluep et al. 2012).

E-Waste exports have been a subject of many discussions at the Basel Convention on the Control of Transboundary Movements of Hazardous Wastes and their Disposal, an UN-treaty signed by 190 countries. The goal of the Basel Convention is to prevent dumping of toxic wastes from industrialized in developing countries.

With the definitions provided (Annex III, List A, subject A1180), transboundary movements of E-Waste should be subject to the rules of the Basel Convention. The Convention requests that importing (and transit-) countries should be notified prior to issuing of an export license of the exporting country. With the BAN Amendment, now ratified by many parties of the Basel Convention, any shipment of hazardous waste from countries of the OECD, the EU, and Lichtenstein to countries not listed in Annex IV of the convention is not allowed.

However, there has been a loophole used by scruples actors: Equipment, declared as "Products for Re-Use or Repair" are by definition not waste and therefore not subject to the rules of the Basel Convention. As there are significant amounts of products shipped for this purpose abroad (e.g., to remanufacturers in Asia), customs are not able to control the shipments if they may contain E-Waste instead of products shipped for repair/remanufacturing. Customs officers were missing also clear criteria to distinguish between products which are shipped for repair/remanufacturing or E-Waste using this loophole.

Starting in 2012, the parties of the Basel Convention tried to solve this issue. In 2019, they agreed "on a temporary basis" Technical Guidelines (UNEP/CHW.14/7/Add.6/Rev.1.) which should provide clarity on the distinction between waste and products shipped for re-use, repair, refurbishment and remanufacturing and remanufacturing.

Discussions on this document, especially on section 32(a) and 32(b) are still going on.

To what extent are the rules of the Basel Convention impacting local E-Waste solutions?

Rules of the Basel Convention are not binding to individuals or companies. As any UN-Treaty, parties to the Basel Convention are obliged to transpose these rules into national legislation. Enforcement can only happen as soon as they are transposed into national legislation

> Advice **To avoid any issues with improper waste management, preference should be given to domestic waste treatment.** However, when significant cost differences (e.g., due to lower labor cost) or technical requirements (e.g., certain recycling disposal technology not available in-country) require the export of E-Waste (or residuals of previous treatment steps), the following elements considered:
>
> - Exporters from OECD and EU countries (not to forget Lichtenstein) should only export E-Waste for further treatment to other OECD/EU countries.
> - Exporters should establish in any case regular audits to proof that the waste they exported is treated appropriately (e.g., not exported again).
> - Special attention should be given if PROs separate equipment from the waste stream for re-use. If these products are shipped across borders, they should be fully functional. A test, e.g., printing a test page or that the operating system of a computer loads properly will help to proof to customs to distinguish waste from non-waste.
> - Before shipping products from the waste stream abroad, actors should calculate if the potential risk of being involved in negative PR or even punishment is worthwhile the (most likely) small savings from handling the equipment abroad.

References

Puckett, J., Westervelt, S., Gutierrez, R., & Takamyia, Y. (2005). *The digital dump, exporting re-use and abuse to Africa.* Seattle, WA., USA: The Basel Action Network Waste Act No. 59 of 2008) (2020). Ministery of Foresttry, Fisheries and the Environment, Republik South Africa. Retrieved February 19, 2023, from https://www.gov.za/sites/default/files/gcis_document/202105/44539gon400.pdf.

Schluep, M., et al. (2012). Where are WEEE in Africa. In *Proceedings of Electronics Goes Green 2012+ (ECG 2012), Joint International Conference and Exhibition.*

Appendix A

I. **Pricing Models and PRO Market Structure in Countries**

This overview is based on research from 2021. Countries may have changed their WEEE-Systems since then. It is recommended that in case information from these tables are used in other research, to check the actual situation.

Table A.1 Europe, Middle East and Africa

Country	Pricing model	Multiple PROs (IT)	Country	Pricing model	Multiple PROs (IT)
Austria	ACT	Yes	Latvia	ACT	Yes
Belgium	ARF	No	Lithuania	ACT	Yes
Bulgaria	ACT	Yes	Netherlands	ACT	Yes
Czech Republic	ACT	Yes	Norway	ACT	Yes
Denmark	ACT	Yes	Poland	ACT	Yes
Estonia	ACT	Yes	Portugal	ACT	Yes
Finland	ACT	Yes	Romania	ACT	Yes
France	ARF	Yes (only 2)	Slovakia	ACT	Yes
Germany	ACT	Yes	Slovenia	ACT	Yes
Greece	ACT	Yes	South Africa	ARF	Yes
Hungary	ARF	No (gov agency)	Spain	ACT	Yes
Ireland	ACT	Yes (only 2)	Sweden	ACT	Yes
Israel	ACT	Yes	Switzerland	ARF	No
Italy	ACT	Yes	Turkey	ACT	Yes
Croatia	ARF (tax)	No	UK	ACT	Yes

K. Hieronymi, *Waste from Electrical and Electronic Equipment (WEEE) in Practice*, https://doi.org/10.1007/978-3-031-40722-2

Table A.2 Americas

Country	Pricing model	Multiple PROs (IT)	Country	Pricing model	Multiple PROs (IT)
USA			*Canada*		
California	ARF	No	Ontario	ACT	Yes
Connecticut	ACT	Yes	British Colombia	ARF	No
DC	ACT	Yes	Yukon	ARF	No
Hawaii	ACT	Yes	Northwest Territories	ARF	No
Illinois	ACT	Yes	Alberta	ARF	No
Indiana	ACT	Yes	Saskatchewan	ARF	No
Maine	ACT	No	Manitoba	ARF	No
Michigan	ACT ➔ ARF	Yes	Quebec	ARF	No
Minnesota	ACT	Yes	New Brunswick	ARF	No
New Jersey	ARF	Yes	Nova Scotia	ARF	No
New York	ACT	Yes	Prince Edward Island	ARF	No
North Carolina	ACT	Yes	Newfoundland & Labrador	ARF	No
Oklahoma	ACT	Yes	*Latin America*		
Oregon	ACT	No	Argentina	ACT	Yes
Pennsylvania	ACT	Yes	Brazil	ACT	Yes
Rhode Island	ACT	No	Chile	ACT	Yes
South Carolina	ACT	Yes	Colombia	ACT	Yes
Texas	ACT	Yes	Mexico	ACT	Yes
Utah	ACT	Yes	Peru	ACT	Yes
Vermont	ACT	No			
Virginia	ACT	Yes			
Washington	ACT	No			
West Virginia	ACT	Yes			
Wisconsin	ACT	Yes			

Legend:

Pricing Model: **ARF** = Advanced Recycling Fee (Fixed Fees paid for every product sold).

ARF (tax) = Advanced Recycling Fee charged by the government, use of the fee not transparent.

ACT = Amounts Collected/Treated (Producers are assigned a certain amount of Waste).

Multiple PROS: **Yes** = Producers have a choice between multiple PROs.

Table A.3 Asia/Pacific

Country	Pricing model	Multiple PROs (IT)
Australia	ARF	Yes
China	ARF (tax)	No
Hong Kong	ARF	No
India	ACT	Yes
Japan	ARF (IPR)	Yes
Korea	ARF	No
Singapore	ARF	No
Taiwan	ARF	No (gov agency)
Vietnam	?	?

No = Producer must join the only existing PRO.

No (gov agency)= Operation managed by Government Agency or an organization close to the government.

Glossary

Expression	Abbreviation	Short description
% of POM	POM	Collections in relation to amounts Put on the Market
Advanced Recycling Fees	ARF	Charging the Producer based on a fixed fee per product sold
Amounts Collected/ Treated	ACT	Charging the Producer based on the Amounts Collected/Treated by the PRO
Cannibalization		Products are taken appart and parts are used as spare parts to repair other products
Cathode Ray Tube	CRT	TVs and Monitors based on "Cathode Ray Tube" technology. The tubes are made out of leaded glass and contain a phosphorous coating. As better technologies (e.g., Flatscreen) are available, there is currently no significant re-use of CRT-based products and materials
Collection Site		Sites where customers can drop their unwanted equipment. Collections Sites can be set up by municipalities, dealers, recyclers, or PROs
Collective Producer Responsibility	CPR	All Producers are responsible for all waste of their product category
Collection Category		Ideally, every product type would be collected separately to get the maximum impact of EPR. In practice, similar products have are collected together in "Collection Categories". As the composition of Collection Categories are impacting the cost of individual OEMs and the achievement of Collection Rates, attention has to be paid on the composition of Collection Categories
Competent Authority (Basel Convention)		A competent authority means the governmental authority designated to be responsible, within such geographical areas as the Party may think fit, for receiving notifications of transboundary movements and any related information and for responding to such notifications (Article 2)
Convenience Target	CT	100% of E-Waste provided at (convenient) Collection sites must be collected and treated

(continued)

K. Hieronymi, *Waste from Electrical and Electronic Equipment (WEEE) in Practice*, https://doi.org/10.1007/978-3-031-40722-2

(continued)

Expression	Abbreviation	Short description
Design for Recycling	DFR	Improving Recycling ability, e.g., by avoiding hazardous materials, easy dismantling, less composite materials
Directive on Waste from Electrical and Electronic Products	WEEE	Directive of the EU to regulate E-Waste management based EPR. First version established in 2004, revised in 2012
End-of-life Management	EOL	Managing the collection, transportation, recycling, and disposal for products at the end of their viable usage
Extended Producer Responsibility	EPR	Makes Producers responsible for e.g., the end-of-use phase of their products
Free Riders		Producers who are not contributing to the management of waste from their products
Generic Target		Generic targets set across all Collection Categories in case scope includes products outside of ICT
Historic Waste		Products, who's Producer left a market before WEEE-Legislation has been established
Individual Collection of Producers		Producers organize their own collection, e.g., *return events*. These activities are on top of their participation in a Collective Take-Back System
Individual Producer Responsibility	IPR	Each Producer is responsible for its individual products only
Informal Sector		Waste Pickers and treaters, very often without waste licenses collecting and treating waste. Major target are metals which can be re-sold. The expression is also used for all collections outside of PROs (e.g., including charity organizations, scrap waste dealers)
International Accounting Standards	IAS	Requiring (Producers) to establish accruals/cash reserves for future waste management obligations caused by today's sales
IT/C specific Target		Specific target(s) set for IT/C products only
Large Domestic Appliances	LDA	Large Household Appliances (e.g., washing machines, dishwasher, dryer). While SDA (Small Domestic Appliances) can be handled by a single person, moving LDA need more people or mechanical assistance (e.g., fork-lift)
Modulated Collection Obligation		Adaption of Modulation in an ACT-based system. Products that fulfill certain criteria will have to collect less. Those who are less sustainable must collect more than their market share requires
Modulated Fees	MF	Products who are fulfilling certain criteria (e.g., easy to recycle, repairability) will pay lower ARFs, those who are less sustainable pay more
Modulation		Rewarding Producers with more sustainable products, punishing those who are less sustainable

(continued)

(continued)

Expression	Abbreviation	Short description
Negotiation between PROs		Adjustments of Over-/Under-Collection negotiated between PROs
Notification (Basel Convention)		Informing the Competent Authorities on the planned import of Waste into their country and requesting approval
Original Equipment Manufacturer	OEM	A company who designs products and sell them under their brand. Producers (within the WEEE Context) are companies who put products on the local market. They can be OEMs who produce locally, OEMs who import, or distributors/whole sellers who import products without being the OEM
Orphan Products		Products where the Producer is no longer active in a jurisdiction (e.g., due to bankruptcy)
Pre-Consent (Basel Convention)		Facilities may receive a generic approval to receive certain wastes, limited for a certain time. A *Pre-Consent* document avoids a specific approval for a single shipment
Pricing Category		A group of products who are charged the same ARF, e.g., Printers/Notebooks. There are also *sub-categories* established on weight (e.g., printers weight<4 kg, 4–7 kg, 7–10 kg) or size (screen<12″, 12–17″, 17.20″….)
Pricing Group		Metonym to Pricing Category
Polluter Pays Principle	PPP	Polluters are responsible for the impact of their products on the environment. In case of E-Waste, Producers are the polluter and therefore responsible for the waste from their products
Producer Responsibility Organization	PRO	An organization to whom Producers are subcontracting the operation of their compliance obligations (e.g., pick-up, transportation, recycling, and disposal)
Producers		Companies who place products on a market, e.g., by manufacturing in-country or importing. Can be OEMs but also importers / channel partners. Depend on who the legal importer is
Product Groups		Product of a similar type/use (e.g., Printer, Notebook), very often combined in "Product Categories", "Pricing Categories"
Put On the Market	POM	Products sold in a jurisdiction within a period
Recovery		Recovering the energy in a material, e.g., by using it as a fuel in a power plant
Recycling		Turning waste into materials which can be used either for the same purpose or in different applications
Restriction of Hazardous Substances	ROHS	EU-Regulation which restricts the use of certain hazardous substances

(continued)

(continued)

Expression	Abbreviation	Short description
Return of Investment	ROI	Profitability of an investment. In context of WEEE, pay off from investments if DFR, but also advocacy elements
Re-Use		The expression Re-Use in the sense of the WEEE-Directive is limited to products picked out of the waste and used again. Products sold on the 2nd hand market (without being discarded by the previous owner) are not considered "Re-Use"
Social Enterprise		Enterprises whose major goal is not to earn profits but to provide jobs, for example, the handicapped or long-time unemployed people. They are very often subsidized and can offer their services at low prices
Small Domestic Appliances	SDA	Small appliances (e.g., blenders, vacuum cleaners) used in households. While SDA can be handled by a single person, Large Domestic Appliances (LDA) more people or mechanical assistance (e.g., fork-lift) are required to move them
Take-Back Regulation		Legislation based on EPR who require Producers to organize and/or pay for the collection and treatment of products. Established for many product groups, e.g., electronic equipment, packaging, batteries
Take-Back System	TBS	Operational implementation structure for E-Waste solutions based on EPR
Treatment		Treatment includes operations required for proper handling of the waste, including sorting, dismantling, recycling, incineration, disposal
Visible Fee	VF	ARFs visible to the customer as requested by legislation

Bibliography

Annex VIII. Basel convention. Retrieved May, 2023, from http://www.basel.int/TheConvention/Overview/TextoftheConvention/tabid/1275/Default.aspx.

Appliance list. Retrieved April, 2023, from https://www.recupel.be/en/appliance-list/.

Bhutta, M. K. S., Omar, A., & Yang, X. (2011). Electronic waste: A growing concern in today's environment. *Economics Research International, 2011.*

EAR Herstellerverzeichnis. Retrieved May, 2023, from https://www.ear-system.de/ear-verzeichnis/hersteller.jsf#no-back.

European Commission. (2011). Restriction on the use of certain hazardous substances in electrical and electronic equipment. Retrieved February 16, 2023, from https://eur-lex.europa.eu/legal-content/DE/TXT/?uri=celex%3A32011L0065.

European Parliament, Environment policy: general principles and basic framework. Retrieved February 10, 2023, from https://www.europarl.europa.eu/factsheets/en/sheet/71/environment-policy-general-principles-and-basic-framework.

European Parliament and Council: Directive 2012/19/EU (WEEE). Retrieved February 10, 2023, from https://eur-lex.europa.eu/legal-content/EN/TXT/?uri=CELEX%3A02012L0019-20180704.

Fachbericht. (2022). Retrieved April, 2023, from https://www.swico.ch/media/filer_public/ab/78/ab78f423-ab16-4dd1-8d75-3623fe8545a/fachbericht-2022-sens-swico-de.pdf.

Fischer, T. (2018). Discussion of practical implementations from the point of view of a producer responsibility organization (PRO/EPR). Retrieved February 10, 2023, from https://erp-recycling.org/wp-content/uploads/2018/06/ERP-Background-Paper-Modulated-Fees-June-2018.pdf.

Huisman, J., van der Maesen, M., Eijsbouts, R. J. J., Wang, F., Baldé, C. P., & Wielenga, C. A. (2012). The Dutch WEEE flows. *United Nations University, ISP–SCYCLE, Bonn, Germany, March 15, 2012.*

Konventionsunterzeichner SWICO. Retrieved May, 2023, from https://www.swico.ch/de/recycling/entsorgen-und-rezyklieren/konventionsunterzeichner/.

Li, J., et al. (2013). Regional or global WEEE recycling. Where to go? *Waste Management 33*(4), 923–934.

Seyring, N., Kling, M., Weissenbacher, J., Hestin, M., Lecerf, L., Magalini, F., et al. (2015). Study on WEEE recovery targets, preparation for re-use targets and on the method for calculation of the recovery targets. *EU Commission.*

Standard AS/NZ 5377. Retrieved May, 2023, from https://www.standards.org.au/standards-catalogue/sa-snz/publicsafety/ev-019/as-slash-nzs--5377-2013.

UNEP/CHW.14/7/Add.6/Rev.1. Technical guidelines on transboundary movements of electrical and electronic waste and used electrical and electronic equipment, in particular regarding the distinction between waste and non-waste under the Basel Convention, 2019. Retrieved May, 2023, from http://www.basel.int/Implementation/TechnicalMatters/DevelopmentofTechnicalGuidelines/TechnicalGuidelines/tabid/8025/Default.aspx.

K. Hieronymi, *Waste from Electrical and Electronic Equipment (WEEE) in Practice*, https://doi.org/10.1007/978-3-031-40722-2

Walter, L. M., & Skouse, C. (2013). *Liabilities and equity*. https://openlibrary.org/books/OL25881085M/Liabilities_and_Equity.

Walther, L. M. (2023). Contingent liabilities. Retrieved February 18, 2023, from https://www.principlesofaccounting.com/chapter-12/. (Chapter 12).

SPRINGER NATURE

GPSR Compliance

The European Union's (EU) General Product Safety Regulation (GPSR) is a set of rules that requires consumer products to be safe and our obligations to ensure this.

If you have any concerns about our products, you can contact us on ProductSafety@springernature.com

In case Publisher is established outside the EU, the EU authorized representative is:

Springer Nature Customer Service Center GmbH
Europaplatz 3
69115 Heidelberg, Germany

The manufacturer's authorised representative in the EU is Springer Nature Customer Service Centre GmbH, Europaplatz 3, 69115 Heidelberg, Germany. If you have any concerns regarding our products, please contact ProductSafety@springernature.com

Printed and bound by CPI Group (UK) Ltd, Croydon, CR0 4YY
01/10/2026
02225882-0005